QC 174.45 LAB

TOPOLOGICAL QUANTUM FIELD THEORY AND FOUR MANIFOLDS

MATHEMATICAL PHYSICS STUDIES

VOLUME 25

Topological Quantum Field Theory and Four Manifolds

by

JOSE LABASTIDA

and

MARCOS MARINO

 Springer

A C.I.P. Catalogue record for this book is available from the Library of Congress.

ISBN 1-4020-3058-4 (HB)
ISBN 1-4020-3177-7 (e-book)

Published by Springer,
P.O. Box 17, 3300 AA Dordrecht, The Netherlands.

Sold and distributed in North, Central and South America
by Springer,
101 Philip Drive, Norwell, MA 02061, U.S.A.

In all other countries, sold and distributed
by Springer,
P.O. Box 322, 3300 AH Dordrecht, The Netherlands.

Printed on acid-free paper

Printed in the Netherlands.

Table of Contents

Preface vii

1. Topological Aspects of Four-Manifolds 1
   1.1. Homology and cohomology 1
   1.2. The intersection form 2
   1.3. Self-dual and anti-self-dual forms 4
   1.4. Characteristic classes 5
   1.5. Examples of four-manifolds. Complex surfaces 6
   1.6. Spin and Spin$_c$-structures on four-manifolds 9

2. The Theory of Donaldson Invariants 12
   2.1. Yang–Mills theory on a four-manifold 12
   2.2. $SU(2)$ and $SO(3)$ bundles 14
   2.3. ASD connections 16
   2.4. Reducible connections 18
   2.5. A local model for the moduli space 19
   2.6. Donaldson invariants 22
   2.7. Metric dependence 27

3. The Theory of Seiberg–Witten Invariants 31
   3.1. The Seiberg–Witten equations 31
   3.2. The Seiberg–Witten invariants 32
   3.3. Metric dependence 36

4. Supersymmetry in Four Dimensions 39
   4.1. The supersymmetry algebra 39
   4.2. $\mathcal{N} = 1$ superspace and superfields 40
   4.3. $\mathcal{N} = 1$ supersymmetric Yang–Mills theories 45
   4.4. $\mathcal{N} = 2$ supersymmetric Yang–Mills theories 50
   4.5. $\mathcal{N} = 2$ supersymmetric hypermultiplets 53
   4.6. $\mathcal{N} = 2$ supersymmetric Yang–Mills theories with matter . . . 55

5. Topological Quantum Field Theories in Four Dimensions 58
   5.1. Basic properties of topological quantum field theories 58
   5.2. Twist of $\mathcal{N} = 2$ supersymmetry 61
   5.3. Donaldson–Witten theory 64
   5.4. Twisted $\mathcal{N} = 2$ supersymmetric hypermultiplet 71
   5.5. Extensions of Donaldson–Witten theory 72
   5.6. Monopole equations 74

6. The Mathai–Quillen Formalism 78

6.1. Equivariant cohomology 79
6.2. The finite-dimensional case 82
6.3. A detailed example 88
6.4. Mathai–Quillen formalism: Infinite-dimensional case 93
6.5. The Mathai–Quillen formalism for theories with gauge symmetry 102
6.6. Donaldson–Witten theory in the Mathai–Quillen formalism . 105
6.7. Abelian monopoles in the Mathai–Quillen formalism 107

7. The Seiberg–Witten Solution of $\mathcal{N} = 2$ SUSY Yang–Mills Theory   110
   7.1. Low energy effective action: semi-classical aspects 110
   7.2. Sl(2, **Z**) duality of the effective action 116
   7.3. Elliptic curves 120
   7.4. The exact solution of Seiberg and Witten 123
   7.5. The Seiberg–Witten solution in terms of modular forms . . . 129

8. The u-plane Integral 133
   8.1. The basic principle (or, 'Coulomb + Higgs=Donaldson') . . 133
   8.2. Effective topological quantum field theory on the u-plane . . 134
   8.3. Zero modes 140
   8.4. Final form for the u-plane integral 144
   8.5. Behavior under monodromy and duality 149

9. Some Applications of the u-plane Integral 154
   9.1. Wall crossing 154
   9.2. The Seiberg–Witten contribution 157
   9.3. The blow-up formula 165

10. Further Developments in Donaldson–Witten Theory 170
    10.1. More formulae for Donaldson invariants 170
    10.2. Applications to the geography of four-manifolds 177
    10.3. Extensions to higher rank gauge groups 188

Appendix A. Spinors in Four Dimensions 204

Appendix B. Elliptic Functions and Modular Forms 209

Bibliography 213

Preface

The emergence of topological quantum field theory has been one of the most important breakthroughs which have occurred in the context of mathematical physics in the last century, a century characterized by independent developments of the main ideas in both disciplines, physics and mathematics, which has concluded with two decades of strong interaction between them, where physics, as in previous centuries, has acted as a source of new mathematics. Topological quantum field theories constitute the core of these phenomena, although the main driving force behind it has been the enormous effort made in theoretical particle physics to understand string theory as a theory able to unify the four fundamental interactions observed in nature. These theories set up a new realm where both disciplines profit from each other. Although the most striking results have appeared on the mathematical side, theoretical physics has clearly also benefitted, since the corresponding developments have helped better to understand aspects of the fundamentals of field and string theory.

Topology has played an important role in the study of quantum mechanics since the late fifties, after discovering that global effects are important in physical phenomena. Many aspects of topology have become ordinary elements in studies in quantum mechanics as well as in quantum field theory and in string theory. The origin of topological quantum field theory can be traced back to 1982, although the term itself appeared for the first time in 1987. In 1982 E. Witten studied supersymmetric quantum mechanics and supersymmetric sigma models providing a framework that led to a generalization of Morse theory. This framework was later considered by A. Floer who constructed its mathematical setting and enlarged it to a more general context. This, in turn, was reconsidered by E. Witten who, influenced by M. Atiyah, proposed the first topological quantum field theory itself in 1987. His construction consisted of a quantum field theory representation of the theory of Donaldson invariants on four-manifolds proposed in 1982.

After the first formulation of a topological quantum field theory by E. Witten many others have been considered. A new area of active research has developed since then. In this book we will concentrate our attention on aspects related to that first theory, nowadays known as Donaldson–Witten theory, which is the most relevant theory in four dimensions. Other important topological theories, such as Chern–Simons gauge theory in three dimensions or topological string theory, fall beyond the scope of this book. We will deal with many aspects of Donaldson–Witten theory, emphasizing how its formulation has allowed Donaldson invariants to be expressed in terms of a set of new simpler invariants known as Seiberg–Witten invariants.

Topological quantum field theory is responsible for the discovery of Seiberg–Witten invariants and their relation to Donaldson invariants. In general, quantum field theories can be studied by different methods providing several pictures of the same theory. The relation between Donaldson–Witten theory and Donaldson invariants was found using perturbative methods in the context of quantum field theory. The application of non-pertubative methods to the same theory waited several years but led to the discovery of the relevance of Seiberg–Witten invariants as building blocks of Donaldson invariants. This connection was possible owing to the progress achieved in 1994 by N. Seiberg and E. Witten in understanding non-perturbative properties of supersymmetric theories.

From the mathematical side the emergence of Seiberg–Witten invariants constitutes one of the most important results obtained in the nineties in the study of four-manifolds. These invariants turn out to be much simpler than Donaldson invariants and contain all the information provided by the latter. To understand the connection between these invariants one needs to regard Donaldson–Witten theory as a theory which originates after the twisting of certain supersymmetric quantum field theories. Other pictures of topological quantum field theory, such as the one in the framework of the Mathai–Quillen formalism, which is also described in this book, do not provide useful information in this respect. However, it is important to become acquainted with this picture since it provides an interesting geometrical setting.

The main goal of this book is to provide a unifying treatment of all the aspects of Donaldson–Witten theory as a stem theory for Donaldson and Seiberg–Witten invariants. An important effort has been made so that it can be read by theoretical physicists and mathematicians. The focus of the book is on the interplay of mathematical and physical aspects of the theory, and although we have included expositions of background material —such as the more mathematical aspects of Donaldson theory or the physics of the

Seiberg–Witten solution— we have not provided all the details, and we refer the reader to more specific references for an exhaustive treatment of some of the subjects.

The book starts with a chapter that collects basic mathematical results about the topology of four-manifolds which will be needed in the rest of the chapters. Chapters 2 and 3 contain reviews of the theories of Donaldson and Seiberg–Witten invariants, respectively. Chapter 4 presents supersymmetry in four dimensions and describes the supersymmetric theories which will be relevant for Donaldson–Witten theory. Chapter 5 deals with the twisting of supersymmetric theories and constructs all the topological quantum field theories which will be of interest in other chapters. There is shown, in particular, in sections 5.3 and 5.6, the connection between these theories and the Donaldson and Seiberg–Witten invariants introduced in Chapters 2 and 3. In Chapter 6 a different framework for dealing with topological quantum field theories, the Mathai–Quillen formalism, is introduced. This formalism provides an interesting geometrical framework for these theories which is worth being be considered. However, its content is not needed for the rest of the book. The chapter could be omitted in a first reading. Chapter 7 deals with non-perturbative aspects of supersymmetric theories. A detailed analysis of the resulting solution, the Seiberg–Witten solution, is presented. The structure of this solution is used in Donaldson–Witten theory in Chapter 8. It allows one to write the Donaldson–Witten invariants as an integral on the so called u-plane introduced by Moore and Witten. The u-plane integral is the most systematic physical framework in which to understand Donaldson–Witten theory, and it also leads to the connection between Donaldson invariants and Seiberg–Witten invariants. Chapter 9 deals with several applications of the u-plane integral, and Chapter 10 summarizes further developments of Donaldson–Witten theory. Finally, two appendices contain useful formulae about spinors in four dimensions, elliptic functions and modular forms.

Acknowledgements

We would like to thank our collaborators and colleagues over all the years we have devoted to the study of topological quantum field theory. We have benefitted from their knowledge and their insight has certainly influenced our work. It is not possible to list all of them here but we wish to thank specially M. Alvarez, L. Álvarez-Gaumé, J.D. Edelstein, C. Lozano, J. Mas, G. Moore, M. Pernici, A.V. Ramallo, and E. Witten.

Several colleagues agreed to read part of the manuscript before publication, providing important remarks. We would like to give special thanks in this respect to Carlos Lozano, Gregory Moore, and Vicente Muñoz.

Chapter 1

Topological Aspects of Four-Manifolds

The purpose of this chapter is to collect a series of basic results about the topology of four-manifolds that will be used in the rest of the book. No attempt to be self-contained is made and the reader should consult some of the excellent books on the subject reviewed at the end of the chapter. The discussion will be restricted to four-manifolds which are closed, compact and orientable, which is the case considered in the rest of the book. We will also assume that all the four-manifolds under consideration are endowed with a Riemannian metric.

1.1. Homology and cohomology

The most basic classical topological invariants of a four-manifold are the homology and cohomology groups $H_i(X, \mathbf{Z})$, $H^i(X, \mathbf{Z})$. These homology groups are abelian groups, and the rank of $H_i(X, \mathbf{Z})$ is called the i-th *Betti number* of X, denoted by b_i. Remember that, for an n-dimensional manifold, by Poincaré duality, one has

$$H^i(X, \mathbf{Z}) \simeq H_{n-i}(X, \mathbf{Z}), \tag{1.1}$$

and also $b_i = b_{n-i}$. We will also need the (co)homology groups with coefficients in other groups such as $\mathbf{Z}_2$. To obtain these groups one uses the universal coefficient theorem, which states that

$$H_i(X, G) \simeq H_i(X, \mathbf{Z}) \otimes_{\mathbf{Z}} G \oplus \mathrm{Tor}(H_{i-1}(X, \mathbf{Z}), G). \tag{1.2}$$

Let us focus on the case $G = \mathbf{Z}_p$. Given an element x in $H_i(X, \mathbf{Z})$, one can always find an element in $H_i(X, \mathbf{Z}_p)$ by sending $x \to x \otimes 1$. This in fact gives a map:

$$H_i(X, \mathbf{Z}) \to H_i(X, \mathbf{Z}_p) \tag{1.3}$$

1

which is called the *reduction* mod p of the class x. Notice that, by construction, the image of (1.3) is in $H_i(X, \mathbf{Z}) \otimes \mathbf{Z}_p$. Therefore, if the torsion part in (1.2) is not zero, the map (1.3) is clearly not surjective. When the torsion part is zero, any element in $H_i(X, \mathbf{Z}_p)$ comes from the reduction mod p of an element in $H_i(X, \mathbf{Z})$. For the cohomology groups we have a similar result.

Physicists are more familiar with the *de Rham cohomology groups*, $H_{DR}^*(X)$ which are defined in terms of differential forms. These groups are defined over $\mathbf{R}$, and therefore they are insensitive to the torsion part of the singular cohomology. Formally, one has $H_{DR}^i(X) \simeq (H^i(X, \mathbf{Z})/\mathrm{Tor}(H^i(X, \mathbf{Z}))) \otimes_{\mathbf{Z}} \mathbf{R}$.

Remember also that there is a non-degenerate pairing in cohomology, which in the de Rham case is the usual wedge product followed by integration. We will denote the pairing of the cohomology classes (or differential form representatives) α, β by (α, β).

Let us now focus on dimension four. Poincaré duality then gives an isomorphism between $H_2(X, \mathbf{Z})$ and $H^2(X, \mathbf{Z})$. It also follows that $b_1(X) = b_3(X)$. Recall that the Euler characteristic $\chi(X)$ of an n-dimensional manifold is defined as

$$\chi(X) = \sum_{i=0}^{n} (-1)^i b_i(X). \tag{1.4}$$

For a connected four-manifold X, we then have, using Poincaré duality, that

$$\chi(X) = 2 - 2b_1(X) + b_2(X). \tag{1.5}$$

1.2. The intersection form

An important object in the geometry and topology of four-manifolds is the *intersection form*,

$$Q : H^2(X, \mathbf{Z}) \times H^2(X, \mathbf{Z}) \to \mathbf{Z}, \tag{1.6}$$

which is just the pairing restricted to the two-classes. By Poincaré duality it can be defined on $H_2(X, \mathbf{Z}) \times H_2(X, \mathbf{Z})$ as well. Notice that Q is zero if any of the arguments is a torsion element, therefore one can define Q on the torsion-free parts of homology and cohomology.

Another useful way of looking at the intersection form is precisely in terms of the intersection of submanifolds in X. One fundamental result in this respect is that we can represent any two-homology class in a four-manifold by a closed oriented surface S: given an embedding

$$i : S \hookrightarrow X \qquad (1.7)$$

we have a two-homology class $i_*([S]) \in H_2(X, \mathbf{Z})$, where $[S]$ is the fundamental class of S. Conversely, any $a \in H_2(X, \mathbf{Z})$ can be represented in this way, and $a = [S_a]$. One can also prove that

$$Q(a, b) = S_a \cup S_b, \qquad (1.8)$$

where the right hand side is the number of points in the intersection of the two surfaces, counted with signs which depend on the relative orientation of the surfaces. If, moreover, η_{S_a}, η_{S_b} denote the Poincaré duals of the submanifolds S_a, S_b one has

$$Q(a, b) = \int_X \eta_{S_a} \wedge \eta_{S_b} = Q([\eta_{S_a}], [\eta_{S_b}]). \qquad (1.9)$$

If we choose a basis $\{a_i\}_{i=1,\dots,b_2(X)}$ for the torsion-free part of $H_2(X, \mathbf{Z})$ we can represent Q by a matrix with integer entries that we will also denote by Q. Under a change of basis, we obtain another matrix $Q \to C^{\mathrm{T}} Q C$, where C is the transformation matrix. The matrix Q is obviously symmetric, and it follows by Poincaré duality that it is unimodular, $i.e.$, it has $\det(Q) = \pm 1$. If we consider the intersection form on the real vector space $H_2(X, \mathbf{R})$ we see that it is a symmetric, bilinear, non-degenerate form, and therefore it is classified by its rank and its signature. The rank of Q, $\mathrm{rk}(Q)$, is clearly given by $b_2(X)$, the second Betti number. The number of positive and negative eigenvalues of Q will be denoted by $b_2^+(X)$, $b_2^-(X)$, respectively, and the $signature$ of the manifold X is then defined as

$$\sigma(X) = b_2^+(X) - b_2^-(X). \qquad (1.10)$$

We will say that the intersection form is $even$ if $Q(a, a) \equiv 0 \bmod 2$. Otherwise it is odd. An element x of $H_2(X, \mathbf{Z})/\mathrm{Tor}(H_2(X, \mathbf{Z}))$ is called $characteristic$ if

$$Q(x, a) \equiv Q(a, a) \bmod 2 \qquad (1.11)$$

for any $a \in H_2(X, \mathbf{Z})/\mathrm{Tor}(H_2(X, \mathbf{Z}))$. An important property of characteristic elements is that

$$Q(x, x) \equiv \sigma(X) \bmod 8. \qquad (1.12)$$

In particular, if Q is even then the signature of the manifold is divisible by 8.

Examples:

(1) The simplest intersection form is:

$$n(1) \oplus m(-1) = \text{diag}(1, \ldots, 1, -1, \ldots, -1), \qquad (1.13)$$

which is odd and has $b_2^+ = n$, $b_2^- = m$.

(2) Another important form is the hyperbolic lattice,

$$H = \begin{pmatrix} 0 & 1 \\ 1 & 0 \end{pmatrix}, \qquad (1.14)$$

which is even and has $b_2^+ = b_2^- = 1$.

(3) Finally, one has the even positive definite form of rank 8

$$E_8 = \begin{pmatrix} 2 & 1 & 0 & 0 & 0 & 0 & 0 & 0 \\ 1 & 2 & 1 & 0 & 0 & 0 & 0 & 0 \\ 0 & 1 & 2 & 1 & 0 & 0 & 0 & 0 \\ 0 & 0 & 1 & 2 & 1 & 0 & 0 & 0 \\ 0 & 0 & 0 & 1 & 2 & 1 & 0 & 1 \\ 0 & 0 & 0 & 0 & 1 & 2 & 1 & 0 \\ 0 & 0 & 0 & 0 & 0 & 1 & 2 & 0 \\ 0 & 0 & 0 & 0 & 1 & 0 & 0 & 2 \end{pmatrix}, \qquad (1.15)$$

which is the Dynkin diagram of the exceptional Lie algebra E_8.

Fortunately, unimodular lattices have been classified. The result depends on whether the intersection form is even or odd and whether it is definite (positive or negative) or not. Odd, indefinite lattices are equivalent to $p(1) \oplus q(-1)$, whilst even indefinite lattices are equivalent to $pH \oplus qE_8$. Definite lattices are more complicated, since they involve 'exotic' cases.

The intersection form is clearly a homotopy invariant. It turns out that simply connected smooth four-manifolds are completely characterized topologically by the intersection form, $i.e.$, two simply connected smooth four-manifolds are homeomorphic if their intersection forms are equivalent. This is a result owed to Freedman. The classification of smooth four-manifolds up to diffeomorphism is another story, and this is the main reason for introducing new invariants which are sensitive to the differentiable structure. But before going into that we have to give some more details about classical topology.

1.3. Self-dual and anti-self-dual forms

The Riemannian structure on the manifold X allows us to introduce the Hodge operator $*$ in de Rham cohomology, that can be used to define an induced metric in the space of forms by

$$d\mu\,(\psi, \theta) = \psi \wedge *\theta, \tag{1.16}$$

where $d\mu$ is the Riemannian volume element. In four dimensions the Hodge operator maps $* : \Omega^2(X) \to \Omega^2(X)$, and since $*^2 = 1$ it has eigenvalues ± 1. It then gives a splitting of the two-forms in $\Omega^2(X)$ in self-dual (SD) and anti-self-dual (ASD) forms, defined as the ± 1 eigenspaces of $*$ and denoted by $\Omega^{2,+}(X)$ and $\Omega^{2,-}(X)$, respectively. Given a differential form ψ, its SD and ASD parts will be denoted by $\psi^\pm$. Explicitly,

$$\psi^\pm = \frac{1}{2}(\psi \pm *\psi). \tag{1.17}$$

The Hodge operator lifts to cohomology and gives a map

$$* : H^2(X) \to H^2(X). \tag{1.18}$$

The number of $+1$ eigenvalues in $H^2(X)$ is precisely $b_2^+(X)$, and the number of -1 eigenvalues is $b_2^-(X)$. We will denote the space of SD and ASD two-cohomology classes as $H^{2,\pm}(X)$, respectively. It follows from this that we can interpret b_2^+ as the number of self-dual harmonic forms on X, and this interpretation will be useful later.

1.4. Characteristic classes

An important set of topological invariants of X is given by the characteristic classes of its real tangent bundle. The most elementary ones are the Pontrjagin class $p(X)$ and the Euler class $e(X)$, both in $H^4(X, \mathbf{Z}) \simeq \mathbf{Z}$. These classes are then completely determined by two integers once a generator of $H^4(X, \mathbf{Z})$ is chosen. These integers will be also denoted by $p(X)$, $e(X)$, and they give the Pontrjagin number and the Euler characteristic of the four-manifold X, so $e(X) = \chi$. The Pontrjagin number is related to the signature of the manifold through the Hirzebruch theorem, which states that:

$$p(X) = 3\sigma(X). \tag{1.19}$$

If a manifold admits an almost complex structure one can define a holomorphic tangent bundle $T^{(1,0)}(X)$. This is a complex bundle of rank $r = \dim(X)$, therefore we can associate with it the Chern character $c(T^{(1,0)}(X))$ which is denoted by $c(X)$. For a four-dimensional manifold one has $c(X) = 1 + c_1(X) + c_2(X)$. Since $c_1(X)$ is a two-form its square can be paired with the fundamental class of the four-manifold. The resulting number can be expressed in terms of the Euler characteristic and the signature as follows:

$$c_1^2(X) = 2\chi(X) + 3\sigma(X). \tag{1.20}$$

Finally, the second Chern class of X is just its Euler class: $c_2(X) = e(X)$. If the almost complex structure is integrable then the manifold X is complex, and it is called a complex surface. Complex surfaces provide many examples in the theory of four-manifolds. Moreover, there is a very beautiful classification of complex surfaces owed to Kodaira, using techniques of algebraic geometry.

There is another set of characteristic classes which is perhaps less known in physics. These are the Stiefel–Whitney classes of real bundles F over X, denoted by $w_i(F)$. They take values in $H^i(X, \mathbf{Z}_2)$. The Stiefel–Whitney classes of a four-manifold X are defined as $w_i(X) = w_i(TX)$. The first Stiefel–Whitney class of a manifold measures its orientability, so we will always have $w_1(X) = 0$. The second Stiefel–Whitney class plays an important role in what follows. This is a two-cohomology class with coefficients in $\mathbf{Z}_2$, and it has three important properties. If the manifold admits an almost complex structure then

$$c_1(X) \equiv w_2(X) \bmod 2, \tag{1.21}$$

i.e., $w_2(X)$ is the reduction mod 2 of the first Chern class of the manifold. This is a general property of $w_2(X)$ for any almost complex manifold. In four dimensions $w_2(X)$ satisfies in addition two other properties: first, it always has an integer lift to an integer class (for example, if the manifold is almost complex then $c_1(X)$ is such a lift). The second property is the Wu formula, which states that

$$(w_2(X), \alpha) = (\alpha, \alpha) \bmod 2, \tag{1.22}$$

for any $\alpha \in H^2(X, \mathbf{Z})$. The left hand side can be interpreted as the pairing of α with the integer lift of $w_2(X)$. A corollary of the Wu formula is that an integer two-cohomology class is characteristic if and only if it is an integer lift of $w_2(X)$.

1.5. Examples of four-manifolds. Complex surfaces

Let us first give some simple examples of four-manifolds:

(1) The simplest example is perhaps the four-sphere $\mathbf{S}^4$. It has $b_1 = b_2 = 0$, and therefore $\chi = 2$, $\sigma = 0$, $Q = 0$.

(2) Next we have the complex projective space $\mathbb{P}^2$. Recall that this is the complex manifold obtained from $\mathbf{C}^3 - \{(0,0,0)\}$ by identifying $z_i \simeq \lambda z_i$, $i = 1, 2, 3$, with $\lambda \neq 0$. $\mathbb{P}^2$ has $b_1 = 0$ and $b_2 = 1$. In fact, the basic two-homology class is the so called class of the hyperplane h, which is given in projective coordinates by $z_1 = 0$. It is not difficult to prove that $h^2 = 1$, so $Q_{\mathbb{P}^2} = (1)$. Notice that h is, in fact, a $\mathbb{P}^1$, therefore it is an embedded sphere in $\mathbb{P}^2$. The projective plane with the opposite orientation will be denoted by $\overline{\mathbb{P}^2}$, and it has $Q = (-1)$.

(3) An easy way to obtain four-manifolds is by taking products of two Riemann surfaces. A simple example is the so called product ruled surfaces $\mathbf{S}^2 \times \Sigma_g$, where Σ_g is a Riemann surface of genus g. This manifold has $b_1 = 2g$, $b_2 = 2$. The homology classes have the submanifold representatives $\mathbf{S}^2$ and Σ_g. They have self-intersection zero and they intersect in one point, therefore $Q = H$, the hyperbolic lattice, with $b_2^+ = b_2^- = 1$. One then has $\chi = 4(1 - g)$.

(4) A more complicated example is the hypersurface of degree d in $\mathbb{P}^3$ described by a homogeneous polynomial $\sum_{i=1}^{4} z_i^d = 0$. We will denote this surface by S_d. It is an algebraic surface, hence a complex manifold. Its first and second Chern numbers can be easily computed by using the adjunction formula, and one finds $c_1^2 = (4 - d)^2 d$, $c_2 = \chi = d(6 - 4d + d^2)$. Using (1.20) one further deduces that $\sigma = \frac{1}{3}(4 - d^2)d$. For $d = 4$, one obtains the so called $K3$ surface which has $\chi = 24$ and $\sigma = -16$. This manifold possesses the following intersection form:

$$Q_{K3} = 3H \oplus 2(-E_8), \tag{1.23}$$

where H and E_8 are given in (1.14) and (1.15), respectively.

(5) Given any four-manifold X one can always form the connected sum $\widehat{X} = X \sharp \overline{\mathbb{P}^2}$, also called the *blow-up* of X. The sphere $\mathbb{P}^1$ corresponding to the line in $\overline{\mathbb{P}^2}$ is called the exceptional sphere of $\widehat{X}$, and its homology class B satisfies $B^2 = -1$. The two-homology of the blow-up is simply given by $H_2(\widehat{X}, \mathbf{Z}) = H_2(X, \mathbf{Z}) \oplus \langle B \rangle$, and the class B is orthogonal to all the classes inherited from X.

An important class of four-manifolds is given by complex surfaces, *i.e.*, complex manifolds of complex dimension two. There is a beautiful classification of these manifolds owed to Kodaira, which we will briefly review in order to present more examples of four-manifolds.

Complex surfaces can be classified according to the Kodaira dimension $\kappa(S)$. This quantity measures the number of holomorphic $(2n, 0)$ forms for large n and can take the values $-\infty, 0, 1$ and 2. This classification is up to bi-rational equivalence, and since blowing-up is a bi-rational transformation it is convenient to focus on minimal surfaces, $i.e.$ surfaces that do not contain holomorphically embedded spheres of square -1 (and therefore they are not the blow-up of another surface). The classification of minimal surfaces then goes as follows:

• The minimal surfaces with $\kappa(S) = -\infty$ are $\mathbb{P}^2$, $ruled$ $surfaces$ ($i.e.$, a sphere bundle over a Riemann surface Σ_g of genus g) or belong to the so called Kodaira class VII. All these surfaces have $b_2^+ = 0$ (for class VII) or 1. For a geometrically ruled surface one has $c_1^2 = 8(1 - g)$, where g is the genus of the base, and they include the products considered in example (3) above. Geometrically ruled surfaces over $\mathbb{P}^1$ are called $Hirzebruch$ $surfaces$, and they are labelled by a non-negative integer n. We will denote them by $\mathbf{F}_n$. One has $\mathbf{F}_0 = \mathbb{P}^1 \times \mathbb{P}^1$ and $\mathbf{F}_1$ is $\mathbb{P}^2$ blown up at one point (therefore it is not minimal).

• There are five types of minimal surfaces with $\kappa(S) = 0$: Enriques surfaces, bi-elliptic surfaces, Kodaira surfaces (primary and secondary), abelian surfaces (tori), and $K3$ surfaces. Enriques, bi-elliptic, and secondary Kodaira surfaces have $b_2^+ \leq 1$. Tori, $K3$, and primary Kodaira surfaces are in fact elliptic fibrations, which we consider now.

• An $elliptic$ $fibration$ is a complex surface S together with a holomorphic fibration $\pi : S \to \Sigma_g$ over a Riemann surface of genus g, where the generic fibers are elliptic curves. All the minimal surfaces with Kodaira dimension $\kappa(S) = 1$ are elliptic fibrations, but the converse is not true, as the examples above show. Any minimal elliptic surface has $c_1^2 = 0$. A particularly interesting sub-family of elliptic surfaces is given by the simply connected four-manifolds $E(n)$, labelled by a non-negative integer n. These manifolds can be constructed starting with $E(1) = \mathbb{P}^2 \sharp 9 \overline{\mathbb{P}^2}$, $i.e.$, the blow-up of $\mathbb{P}^2$ at nine points. It can be shown that this rational surface is elliptic. One can now perform an operation called fiber sum in order to obtain more elliptic fibrations, and in this way one generates the full series $E(n)$. One has $\chi(E(n)) = 12n$ and $\sigma(E(n)) = -8n$, and furthermore $b_2^+(E(n)) = 2n - 1$. $E(2)$ turns out to be a $K3$ surface.

• Finally, we have surfaces with Kodaira dimension $\kappa(S) = 2$, which are called *surfaces of general type*. They have $c_1^2 > 0$.

Of course, complex surfaces are just a small set in the zoo of four-manifolds, and in the latter years many 'exotic' manifolds have been constructed outside the complex realm. The reader is referred to the bibliography at the end of this chapter for a detailed treatment of these developments.

1.6. Spin *and* Spin$_c$-*structures on four-manifolds*

The Spin$_n$ group is a double covering of the orthogonal group $SO(n)$. If E is an oriented $SO(n)$ bundle on a manifold X, a natural question is whether one can lift it to a Spin$_n$-bundle $\mathrm{Sp}(E)$, producing a double covering $\mathrm{Sp}(E) \to E$. If this can be done we say that E is endowed with a *Spin structure*. Locally, Spin structures can always be found, but globally there are topological obstructions which are encoded precisely in the second Stiefel–Whitney class of E, $w_2(E)$. We say that X admits a Spin structure if TX does, and X is then called a *Spin manifold*. The necessary and sufficient condition for X to be Spin is then $w_2(X) = 0$. For example, one can see that the elliptic fibration $E(n)$ introduced in the previous section is Spin if and only if n is even.

If a manifold is Spin then one can consistently construct the spinor bundle S, which is a vector bundle providing a representation of the Clifford algebra. A section of this bundle is nothing but a spinor field. In four dimensions Spin$_4 = SU(2)_+ \times SU(2)_-$, and the spinor bundle decomposes in two irreducible representations $S = S^+ \oplus S^-$, corresponding to positive and negative chirality spinors (our conventions for spinor algebra in Euclidean and Minkowskian signature are collected in Appendix A).

Most four-manifolds are not Spin. In such a situation one can still define something very similar to a spin structure: a Spin$_c$ structure. One of the best ways to think about Spin$_c$ structures is the following: as we said before, the second Stiefel–Whitney cohomology class of a manifold X, $w_2(X) \in H^2(X, \mathbf{Z}_2)$, is always the mod 2 reduction of an integral class w. Let L be the line bundle corresponding to w, *i.e*, $c_1(L) = w$. The square root of the line bundle L in principle does not exist, and the topological obstruction to define it globally is in this case $w_2(X)$, the reduction mod 2 of $c_1(L)$. In the same way, the spinor bundle S of X is not globally defined. However, a standard obstruction analysis in Čech cohomology shows that, although neither $L^{1/2}$ nor S are well defined separately, the product bundle

$$S_L = S \otimes L^{1/2} \qquad (1.24)$$

is well defined. This is easy to understand intuitively: to construct a Spin structure one has to choose a lifting of the rotation group to the spinor group in the different coordinate patches, and as it is well known this lifting involves ± 1 ambiguities. When the choice of signs can not be made consistently for all patches, the Spin structure is obstructed globally. To define the square root of L we have the same kind of ambiguity in signs, but when tensoring both bundles, as in (1.24), these ambiguities compensate each other and the resulting object is globally well defined.

We then see that, given an integer lifting of $w_2(X)$, we can construct a Spin_c structure and define a spinor bundle as in (1.24). Therefore the set of Spin_c structures on a four-manifold is in one-to-one correspondence with the set of characteristic elements. The complex line bundle L is called the determinant line bundle of the Spin_c structure. Once we have found a Spin_c structure we can generate all the rest of them by tensoring with other line bundles: if L_α is the line bundle associated to an element $\alpha \in H^2(X, \mathbf{Z})$, we can construct from (1.24) another spinor bundle

$$S_{L \otimes L_\alpha^2} = S \otimes \left(L^{1/2} \otimes L_\alpha \right). \qquad (1.25)$$

Notice that if $c_1(L) \equiv w_2(X) \bmod 2$ the same is obviously true for $L \otimes L_\alpha^2$. This construction gives a map from $H^2(X, \mathbf{Z})$ to the set of Spin_c structures. If there is no two-torsion in $H^2(X, \mathbf{Z})$ then this map is a bijection, although a non-canonical one since it requires a choice of basepoint (the Spin_c structure associated to L).

In the case of an almost complex manifold X there is, however, a canonical choice of the Spin_c structure: this is for $L = K^{-1}$, where K denotes the canonical line bundle of X. One has, in fact, the isomorphism

$$S_{K^{-1}} = \Omega_{\mathbf{C}}^* T^{(1,0)}(X), \qquad (1.26)$$

where the right hand side denotes the complex exterior powers of the holomorphic tangent bundle of X. Positive chirality spinors correspond to the even powers, and negative chirality spinors to the odd powers.

Since a Spin_c structure can be simply understood as the tensor product (1.24), one can construct the Dirac operator D_L for the Spin_c-structure as the usual Dirac operator coupled to a $U(1)$ connection whose connection and curvature are formally given by $1/2$ the connection and curvature of L. A consequence of (1.12) and the index theorem for the Dirac operator is that if L is the determinant line bundle of a Spin_c-structure then

$$c_1^2(L) = \sigma(X) \bmod 8, \tag{1.27}$$

From this it follows that if X is Spin, then $\sigma(X)$ is a multiple of eight. Actually, it turns out that in fact $\sigma(X)$ is a multiple of 16 for Spin manifolds (Rohlin's theorem).

Bibliographical notes

- A complete survey of the topology of four-manifolds can be found in the excellent book [1]. A short but useful summary is contained in the first chapter of [2].

- The classification of complex surfaces is studied in detail in [3] and [4], as well as in Chapter 3 of [1].

- Hirzebruch surfaces are considered in detail in [3], Chapters III and IV, and the elliptic fibrations $E(n)$ are constructed in detail in [1], Chapter 3. Techniques for constructing 'exotic' four-manifolds are reviewed in [5].

Chapter 2

The Theory of Donaldson Invariants

Donaldson invariants can be mathematically motivated as follows: as we have mentioned, Freedman's results imply that two simply connected smooth manifolds are homeomorphic if and only if they have the same intersection form. However, the classification of four-manifolds up to diffeomorphism turns out to be much more subtle: most of the techniques which one uses in dimension ≥ 5 to approach this problem (like cobordism theory) fail in four dimensions. For example, four dimensions is the only dimension in which a fixed homeomorphism type of closed four-manifolds is represented by infinitely many diffeomorphism types, and $n = 4$ is the only dimension where there are 'exotic' $\mathbf{R}^n$s, *i.e.*, manifolds which are homeomorphic to $\mathbf{R}^n$ but not diffeomorphic to it. One has to look then for a new class of invariants of differentiable manifolds in order to solve the classification problem, and this was the great achievement of Donaldson. Remarkably, the new invariants introduced by Donaldson are defined by looking at instanton configurations of non-abelian gauge theories on the four-manifold. We will give here a sketch of the mathematical procedure for defining Donaldson invariants, in a rather formal way and without entering into the most intricate parts of the theory.

2.1. Yang–Mills theory on a four-manifold

Donaldson theory defines differentiable invariants of smooth four-manifolds starting from Yang–Mills fields on a vector bundle over the manifold. The basic framework is then gauge theory on a four-manifold, and the moduli space of ASD connections.

Let G be a Lie group (usually we will take $G = SO(3)$ or $SU(2)$). Let $P \to X$ be a principal G-bundle over a manifold X with a connection A taking values in the Lie algebra of G, $\mathbf{g}$. Given a vector space V and a representation ρ of G in $GL(V)$, we can form an associated vector bundle $E = P \times_G V$ in the standard way. G acts on V through the representation ρ. The connection A on P induces a connection on the vector bundle E (which we will also denote by A) and a covariant derivative ∇_A. Notice that, whilst the connection A on the principal bundle is an element in $\Omega^1(P, \mathbf{g})$, the induced connection on the vector bundle E is better understood in terms of a local trivialization U_α. On each U_α the connection 1-form A_α is a $gl(V)$ valued one-form (where $gl(V)$ denotes the Lie algebra of $GL(V)$) and the transformation rule which glues together the different descriptions is given by:

$$A_\beta = g_{\alpha\beta}^{-1} A_\alpha g_{\alpha\beta} + i g_{\alpha\beta}^{-1} dg_{\alpha\beta}, \tag{2.1}$$

where $g_{\alpha\beta}$ are the transition functions of E.

Recall that the representation ρ induces a representation of Lie algebras $\rho_* : \mathbf{g} \to gl(V)$. We will identify $\rho_*(\mathbf{g}) = \mathbf{g}$ and define the adjoint action of G on $\rho_*(\mathbf{g})$ through the representation ρ. On X one can consider the *adjoint bundle* $\mathbf{g}_E$, defined by:

$$\mathbf{g}_E = P \times_G \mathbf{g}, \tag{2.2}$$

which is a subbundle of $\mathrm{End}(E)$. For example, for $G = SU(2)$ and V corresponding to the fundamental representation, $\mathbf{g}_E$ consists of Hermitian trace-free endomorphisms of E. If we look at (2.1) we see that the difference of two connections is an element in $\Omega^1(\mathbf{g}_E)$ (the one-forms on X with values in the bundle $\mathbf{g}_E$). Therefore we can think about the space of all connections $\mathcal{A}$ as an affine space whose tangent space at A is given by $T_A\mathcal{A} = \Omega^1(\mathbf{g}_E)$.

The curvature F_A of the vector bundle E associated with the connection A can be also defined in terms of the local trivialization of E. On U_α the curvature F_α is a $gl(V)$-valued two-form that behaves under a change of trivialization as:

$$F_\beta = g_{\alpha\beta}^{-1} F_\alpha g_{\alpha\beta}, \tag{2.3}$$

and this shows that the curvature can be considered as an element in $\Omega^2(\mathbf{g}_E)$.

The next geometrical objects we must introduce are gauge transformations, which are automorphisms of the vector bundle E, $u : E \to E$ preserving the fiber structure (*i.e.*, they map one fiber onto another) and descend to the identity on X. They can be described as sections of the bundle $\mathrm{Aut}(E)$. Gauge transformations form an infinite-dimensional Lie group $\mathcal{G}$, where the group structure is given by pointwise multiplication. The Lie algebra of $\mathcal{G} = \Gamma(\mathrm{Aut}(E))$ is given by $\mathrm{Lie}(\mathcal{G}) = \Omega^0(\mathbf{g}_E)$. This can be seen by looking at the local description, since on an open set U_α the gauge transformation is given by a map $u_\alpha : U_\alpha \to G$, where G acts through the representation ρ. As it is well known, the gauge transformations act on the connections according to

$$u^*(A_\alpha) = u_\alpha A_\alpha u_\alpha^{-1} + idu_\alpha u_\alpha^{-1} = A_\alpha + i(\nabla_A u_\alpha)u_\alpha^{-1}, \qquad (2.4)$$

where

$$\nabla_A u_\alpha = du_\alpha + i[A_\alpha, u_\alpha]. \qquad (2.5)$$

Gauge transformations act on the curvature as:

$$u^*(F_\alpha) = u_\alpha F_\alpha u_\alpha^{-1}. \qquad (2.6)$$

2.2. $SU(2)$ and $SO(3)$ bundles

In this book we will essentially restrict ourselves to the gauge groups $SU(2)$ and $SO(3)$, and the vector bundle will correspond to the fundamental representation of these groups. In the following, $SU(2)$ bundles will be denoted by E, and $SO(3)$ bundles will be denoted by V, so E will be a two-dimensional complex vector bundle and V will be a three-dimensional real vector bundle.

$SU(2)$ bundles over a compact four-manifold are completely classified by the second Chern class $c_2(E)$. In the case of $SO(3)$ bundles, the isomorphism class is completely classified by the first Pontrjagin class

$$p_1(V) = -c_2(V \otimes \mathbf{C}) \qquad (2.7)$$

and the Stiefel–Whitney class $w_2(V) \in H_2(X, \mathbf{Z}_2)$. These characteristic classes are related by

$$w_2(V)^2 = p_1(V) \bmod 4. \qquad (2.8)$$

$SU(2)$ bundles and $SO(3)$ bundles are, of course, related: given an $SU(2)$ bundle we can form an $SO(3)$ bundle by taking the bundle $\mathbf{g}_E$ in (2.2). However, although an $SO(3)$ bundle can be always regarded locally as an $SU(2)$ bundle, there are global obstructions to lift the $SO(3)$ group to an $SU(2)$ group. The obstruction is measured precisely by the second Stiefel–Whitney class $w_2(V)$. Therefore we can view $SU(2)$ bundles as a special case of $SO(3)$ bundles with zero Stiefel–Whitney class, and this is what we are going to do in this book. When the $SO(3)$ bundle can be lifted to an $SU(2)$ bundle one has the relation:

$$p_1(V) = -4c_2(E). \qquad (2.9)$$

Chern–Weil theory gives a representative of the characteristic class $p_1(V)/4$ in terms of the curvature of the connection:

$$\frac{1}{4}p_1(V) = \frac{1}{8\pi^2}\mathrm{Tr}F_A^2, \tag{2.10}$$

where F_A is a Hermitian trace-free matrix valued two-form. Notice that Hermitian trace-free matrices have the form:

$$\xi = \begin{pmatrix} a & -ib+c \\ ib+c & -a \end{pmatrix}, \quad a, b, c \in \mathbf{R}, \tag{2.11}$$

so the trace is a positive definite form:

$$\mathrm{Tr}\,\xi^2 = 2(a^2 + b^2 + c^2) \equiv 2|\xi|^2, \quad \xi \in su(2). \tag{2.12}$$

We define the *instanton number* k as:

$$k = -\frac{1}{8\pi^2}\int_X \mathrm{Tr}F_A^2. \tag{2.13}$$

If V does not have a lifting to an $SU(2)$ bundle, the instanton number is not an integer in general, and satisfies instead

$$k = -\frac{w_2(V)^2}{4} \mod 1. \tag{2.14}$$

If V lifts to E then $k = c_2(E)$.

The topological invariant $w_2(V)$ for $SO(3)$ bundles may be less familiar, but it has been used when $X = \mathbf{T}^4$, the four-torus, to construct gauge configurations called *torons*. Consider $SU(N)$ gauge fields on a four-torus of lengths a_μ, $\mu = 1, \ldots, 4$. Torons are configurations which are topologically non-trivial, and are obtained by requiring the gauge fields to be periodic up to a gauge transformation in two directions:

$$\begin{aligned} A_\mu(a_1, x_2) &= \Omega_1(x_2)A_\mu(0, x_2), \\ A_\mu(x_1, a_2) &= \Omega_2(x_1)A_\mu(x_1, 0), \end{aligned} \tag{2.15}$$

where we have denoted by ΩA the action of the gauge transformation Ω on the connection A. Looking at the corners, we find the compatibility condition

$$\Omega_1(a_2)\Omega_2(0) = \Omega(a_1)\Omega_1(0)Z, \tag{2.16}$$

where $Z \in C(SU(N)) = \mathbf{Z}_N$ is a central element. We can allow a non-trivial Z since a gauge transformation which is in the center of $SU(N)$ does not act on the $SU(N)$ gauge fields. This means that when we allow torons we are effectively dealing with an $SU(N)/\mathbf{Z}_N$ gauge theory. For $SU(2)$ this means that we are dealing with an $SO(3)$ theory, and the toron configurations are, in fact, topologically non-trivial $SO(3)$ gauge fields with non-zero Stiefel–Whitney class.

2.3. ASD connections

The splitting (1.17) between SD and ASD parts of a two-form extends in a natural way to bundle-valued two-forms, and, in particular, to the curvature associated to the connection A, $F_A \in \Omega^2(\mathbf{g}_E)$. We say that a connection is ASD if

$$F_A^+ = 0. \tag{2.17}$$

It is instructive to consider this condition in the case of $X = \mathbf{R}^4$ with the Euclidean metric. If $\{e_1, e_2, e_3, e_4\}$ is an oriented orthonormal frame, a basis for SD (ASD) forms is given by:

$$\{e_1 \wedge e_2 \pm e_3 \wedge e_4, e_1 \wedge e_4 \pm e_2 \wedge e_3, e_1 \wedge e_3 \pm e_4 \wedge e_1\}, \tag{2.18}$$

with $\pm$ for SD and ASD, respectively. If we write $F = \frac{1}{2}F_{\mu\nu}dx^\mu \wedge dx^\nu$ then the ASD condition reads:

$$\begin{aligned} F_{12} + F_{34} &= 0 \\ F_{14} + F_{23} &= 0 \\ F_{13} + F_{42} &= 0. \end{aligned} \tag{2.19}$$

Notice that the second Chern class density can be written as

$$\mathrm{Tr}\,(F_A^2) = \{|F_A^+|^2 - |F_A^-|^2\}d\mu, \tag{2.20}$$

where the norm is defined as

$$|\psi|^2 = \frac{1}{2}\mathrm{Tr}(\psi \wedge *\psi). \tag{2.21}$$

We then see that, with our conventions, if A is an ASD connection the instanton number k is positive. This gives a topological constraint on the existence of ASD connections.

One of the most important properties of ASD connections is that they minimize the Yang–Mills action

$$S_{\mathrm{YM}} = \frac{1}{2}\int_X F \wedge *F = \frac{1}{4}\int_X d^4x \sqrt{g}F_{\mu\nu}F^{\mu\nu} \tag{2.22}$$

in a given topological sector. This is so because the integrand of (2.22) can be written as $|F_A^+|^2 + |F_A^-|^2$, therefore

$$S_{\mathrm{YM}} = \frac{1}{2}\int_X |F_A^+|^2 d\mu + 8\pi^2 k,$$

which is bounded from below by $8\pi^2 k$. The minima are attained precisely when (2.17) holds.

The ASD condition is a non-linear differential equation for non-abelian gauge connections, and it defines a subspace of the infinite-dimensional configuration space of connections A. This subspace can be regarded as the zero locus of the section

$$s : \mathcal{A} \longrightarrow \Omega^{2,+}(\mathbf{g}_E) \tag{2.23}$$

given by

$$s(A) = F_A^+. \tag{2.24}$$

Our main goal is to define a finite-dimensional moduli space starting from $s^{-1}(0)$. The key property to take into account is that section (2.24) is equivariant with respect to the action of the gauge group,

$$s(u^*(A)) = u^*(s(A)). \tag{2.25}$$

Therefore if a gauge connection satisfies the ASD condition then any gauge-transformed connection $u^*(A)$ will be also ASD. By getting rid of this gauge redundancy one obtains a finite-dimensional moduli space, so we must 'divide by $\mathcal{G}$'. In other words, we must quotient $s^{-1}(0)$ by the action of the gauge group. We are thus led to define the moduli space of ASD connections, $\mathcal{M}_{\mathrm{ASD}}$, as follows:

$$\mathcal{M}_{\mathrm{ASD}} = \{[A] \in \mathcal{A}/\mathcal{G} \,|\, s(A) = 0\}, \tag{2.26}$$

where $[\mathcal{A}]$ denotes the gauge equivalence class of the connection $\mathcal{A}$. The above space is well defined since s is $\mathcal{G}$-equivariant. In defining (2.26) it is usual to fix the topological class of the bundle, so one considers only gauge connections in that class. This means, for an $SO(3)$ bundle V, fixing the instanton number k and the second Stiefel–Whitney class $w_2(V)$. When we want to emphasize this property we will write the moduli space as $\mathcal{M}_{\mathrm{ASD}}(w_2(V), k)$.

That the ASD connections form a moduli space is well known in field theory. For example, on $\mathbf{R}^4$ $SU(2)$ instantons are parametrized by a finite number of data (which include, for example, the position of the instanton), giving $8k - 3$ parameters for instanton number k. The moduli space that we have just defined is the generalization of this to an arbitrary, compact four-manifold.

2.4. Reducible connections

In order to analyse $\mathcal{M}_{\mathrm{ASD}}$ we will first look at the map

$$\mathcal{G} \times \mathcal{A} \to \mathcal{A} \qquad (2.27)$$

and the associated quotient space $\mathcal{A}/\mathcal{G}$. The equivalence class of connections in this quotient space are denoted by $[A]$. However it is well known that if the action of the group is not free one has singularities in the quotient space. We then define the isotropy group of a connection A, Γ_A, as

$$\Gamma_A = \{u \in \mathcal{G}|u(A) = A\}, \qquad (2.28)$$

which measures the extent at which the action of $\mathcal{G}$ on A is not free. If the isotropy group is the center of the group $C(G)$ then the action is free and we say that the connection A is *irreducible*. Otherwise we say that the connection A is *reducible*. Reducible connections are well known in field theory, since they correspond to gauge configurations in which the gauge symmetry is broken to a smaller subgroup. For example, the $SU(2)$ connection

$$A = \begin{pmatrix} a & 0 \\ 0 & -a \end{pmatrix} \qquad (2.29)$$

should be regarded as a $U(1)$ connection in disguise. It is clear that a constant gauge transformation of the form $u\sigma_3$ (where $\sigma_3 = \mathrm{diag}(1, -1)$) leaves (2.29) invariant, therefore the isotropy group of A is bigger than the center of $SU(2)$. We will denote the space of irreducible connections by $\mathcal{A}^*$. It follows from the definition that the reduced group of gauge transformations $\widehat{\mathcal{G}} = \mathcal{G}/C(G)$ acts freely on $\mathcal{A}^*$.

By using the description of u as a section of $\mathrm{Aut}(E)$ and the action on A given in (2.4), we see that

$$\Gamma_A = \{u \in \Gamma(\mathrm{Aut}(E))|\nabla_A u = 0\}, \qquad (2.30)$$

i.e., the isotropy group at A is given by the covariantly constant sections of the bundle $\mathrm{Aut}(E)$. It follows that Γ_A is a Lie group, and its Lie algebra is given by

$$\mathrm{Lie}\,(\Gamma_A) = \{f \in \Omega^0(\mathbf{g}_E)|\nabla_A f = 0\}. \qquad (2.31)$$

Therefore, a useful way of detecting whether Γ_A is bigger than $C(G)$ (and has positive dimension) is to study the kernel of ∇_A in $\Omega^0(\mathbf{g}_E)$. Reducible connections then correspond to a non-zero kernel of

$$\nabla_A : \Omega^0(\mathbf{g}_E) \to \Omega^1(\mathbf{g}_E). \qquad (2.32)$$

In the case of $SU(2)$ and $SO(3)$ a reducible connection has precisely the form (2.29), with isotropy group $\Gamma_A/C(G) = U(1)$. This means, topologically, that the $SU(2)$ bundle E splits as

$$E = L \oplus L^{-1}, \tag{2.33}$$

with L a complex line bundle, whilst a reducible $SO(3)$ bundle splits as

$$V = \mathbf{R} \oplus T, \tag{2.34}$$

where $\mathbf{R}$ denotes the trivial rank-one real bundle over X, and T is a complex line bundle. The above structure for V is easily derived by considering the real part of $\mathrm{Sym}^2(E)$. Notice that if V admits a $SU(2)$ lifting E then $T = L^2$. There are topological constraints to have these splittings, because (2.33) implies that $c_2(E) = -c_1(L)^2$, and (2.34) that

$$p_1(V) = c_1(T)^2. \tag{2.35}$$

When E exists the first Chern class $\lambda = c_1(L)$ is an integral cohomology class. However, when $w_2(V) \neq 0$ it then follows from (2.8) that L does not exist as a line bundle, since its first Chern class is not an integral class but lives in the lattice

$$H^2(X, \mathbf{Z}) + \frac{1}{2} w_2(V). \tag{2.36}$$

In particular, one has that

$$c_1(T) \equiv w_2(V) \bmod 2. \tag{2.37}$$

Therefore reductions of V are in one-to-one correspondence with cohomology classes $\alpha \in H^2(X, \mathbf{Z})$ which are congruent to $w_2(V)$ mod 2 and such that $\alpha^2 = p_1(V)$. In the following, when we study the local model of $\mathcal{M}_{\mathrm{ASD}}$ we will restrict ourselves to irreducible connections.

2.5. A local model for the moduli space

To construct a local model for the moduli space means essentially to give a characterization of its tangent space at a given point. The way to do that is to consider the tangent space at an ASD connection A in $\mathcal{A}$, which is isomorphic to $\Omega^1(\mathbf{g}_E)$, and look for the directions in this vector space which preserve the ASD condition and which are not gauge orbits (since we are quotienting by $\mathcal{G}$). The local model for $\mathcal{M}_{\text{ASD}}$ was first obtained by Atiyah, Hitchin, and Singer.

Let us first address the second condition. We want to find out which directions in the tangent space at a connection A are pure gauge, *i.e.*, we want to find slices of the action of the gauge group $\widehat{\mathcal{G}}$. The procedure is simply to consider the derivative of the map (2.27) in the $\mathcal{G}$ variable at a point $A \in \mathcal{A}^*$ to obtain

$$C : \text{Lie}\,(\mathcal{G}) \longrightarrow T_A\mathcal{A}, \tag{2.38}$$

which is nothing but (2.32). Since there is a natural metric in the space $\Omega^*(\mathbf{g}_E)$ we can define a formal adjoint operator:

$$C^\dagger : \Omega^1(\mathbf{g}_E) \longrightarrow \Omega^0(\mathbf{g}_E). \tag{2.39}$$

We can then orthogonally decompose the tangent space at A into the gauge orbit $\text{Im}\,C$ and its complement:

$$\Omega^1(\mathbf{g}_E) = \text{Im}\,C \oplus \text{Ker}\,C^\dagger. \tag{2.40}$$

This is precisely the slice of the action we were looking for. Locally this means that the neighborhood of $[A]$ in $\mathcal{A}^*/\mathcal{G}$ can be modeled by the subspace of $T_A\mathcal{A}$ given by $\text{Ker}\,\nabla_A^\dagger$. Furthermore, the isotropy group Γ_A has a natural action on $\Omega^1(\mathbf{g}_E)$ given by the adjoint multiplication, as in (2.6). If the connection is reducible the moduli space is locally moded on $(\text{Ker}\,\nabla_A^\dagger)/\Gamma_A$.

We have obtained a local model for the orbit space $\mathcal{A}^*/\mathcal{G}$, but we need to enforce the ASD condition in order to obtain a local model for the moduli space of ASD connections modulo gauge transformations. Let A be an irreducible ASD connection, satisfying $F_A^+ = 0$, and let $A + a$ be another ASD connection, where $a \in \Omega^1(\mathbf{g}_E)$. The condition we obtain on a starting from $F_{A+a}^+ = 0$ is $p^+(\nabla_A a + a \wedge a) = 0$, where p^+ is the projector on the SD part of a two-form. At linear order we find:

$$p^+\nabla_A a = 0. \tag{2.41}$$

Note that the map $p^+\nabla_A$ is nothing but the linearization of the section s, ds:

$$ds : T_A\mathcal{A} \longrightarrow \Omega^{2,+}(\mathbf{g}_E). \tag{2.42}$$

The kernel of ds corresponds to tangent vectors that satisfy the ASD condition at linear order (2.41). We can now give a precise description of the tangent space of $\mathcal{M}_{\text{ASD}}$ at A: we want directions which are in $\text{Ker}\, ds$ but which are not in $\text{Im}\, C$. First notice that, since s is gauge equivariant, $\text{Im}\, C \subset \text{Ker}\, ds$. This can be checked by a direct computation

$$p^+ \nabla_A \nabla_A \phi = [F_A^+, \phi] = 0, \quad \phi \in \Omega^0(\mathbf{g}_E), \tag{2.43}$$

since A is ASD. Now taking into account (2.40) we finally find

$$T_{[A]}\mathcal{M}_{\text{ASD}} \simeq (\text{Ker}\, ds) \cap (\text{Ker}\, \nabla_A^\dagger). \tag{2.44}$$

This space can be regarded as the kernel of the operator

$$D : \Omega^1(\mathbf{g}_E) \longrightarrow \Omega^0(\mathbf{g}_E) \oplus \Omega^{2,+}(\mathbf{g}_E) \tag{2.45}$$

given by $D = ds \oplus \nabla_C^\dagger$. Since $\text{Im}\, C \subset \text{Ker}\, ds$ there is a short exact sequence:

$$0 \longrightarrow \Omega^0(\mathbf{g}_E) \xrightarrow{C} \Omega^1(\mathbf{g}_E) \xrightarrow{ds} \Omega^{2,+}(\mathbf{g}_E) \longrightarrow 0. \tag{2.46}$$

This complex is called the *instanton deformation complex* or *Atiyah–Hitchin–Singer (AHS) complex* and gives a very elegant local model for the moduli space of ASD connections. In particular one has that

$$T_{[A]}\mathcal{M}_{\text{ASD}} = H_A^1 \tag{2.47}$$

, where H_A^1 is the middle cohomology group of the complex (2.46):

$$H_A^1 = \frac{\text{ker}\, ds}{\text{Im}\, C}, \tag{2.48}$$

The index of the AHS complex is given by

$$\text{ind} = \dim H_A^1 - \dim H_A^0 - \dim H_A^2, \tag{2.49}$$

where $H_A^0 = \text{Ker}\, C$ and $H_A^2 = \text{Coker}\, ds$. This index is usually called the *virtual* dimension of the moduli space. When A is an irreducible connection (in particular, $\text{Ker}\, \nabla_A = 0$) and, in addition, it satisfies $H_A^2 = 0$ it is then called a *regular* connection. For these connections the dimension of the moduli space is given by the virtual dimension. The index of the AHS complex can be computed for any gauge group G using the Atiyah–Singer index theorem. The result for the group $SO(3)$ is:

$$\dim \mathcal{M}_{\text{ASD}} = -2p_1(V) - \frac{3}{2}(\chi + \sigma). \tag{2.50}$$

The conclusion of this analysis is that if A is an irreducible ASD connection the moduli space in a neighborhood of this point is smooth and can be modeled by the cohomology (2.48). If the connection is also regular, the index of the instanton deformation complex gives the dimension of the moduli space. Of course, the most difficult part of Donaldson theory is to find the global structure of $\mathcal{M}_{\text{ASD}}$. In particular, in order to define the invariants one has to compactify the moduli space. We are not going to deal with these subtle issues here, and refer the reader to the references mentioned at the end of this chapter.

2.6. Donaldson invariants

Donaldson invariants are roughly defined in terms of integrals of differential forms in the moduli space of irreducible ASD connections. These differential forms come from the rational cohomology ring of $\mathcal{A}^*/\mathcal{G} = \mathcal{B}^*$, and it is necessary to have an explicit description of this ring. The construction involves the *universal bundle* or *universal instanton* associated with this moduli problem, and proceeds as follows. For the group $SU(2)$ we will consider the $SO(3)$ bundle $\mathbf{g}_E$, and if the gauge group is $SO(3)$ we consider the vector bundle V. We will denote both of them by $\mathbf{g}_E$, since the construction is the same in both cases. We first consider the space $\mathcal{A}^* \times \mathbf{g}_E$. This space can be regarded as a bundle:

$$\mathcal{A}^* \times \mathbf{g}_E \to \mathcal{A}^* \times X \tag{2.51}$$

which is the pullback from the bundle $\pi : \mathbf{g}_E \to X$. In this construction the space $\mathcal{A}^* \times \mathbf{g}_E$ is called a *family of tautological connections* since the natural connection on $\mathcal{A}^* \times \mathbf{g}_E$ is tautological in the $\mathbf{g}_E$ direction and trivial in the $\mathcal{A}^*$ direction: at the point (A, p) the connection is given by $A_\alpha(\pi(p))$ (where we have chosen a trivialization $\{U_\alpha\}$, and $\pi(p) \in U_\alpha$). Since the group of reduced gauge transformations $\widehat{\mathcal{G}}$ acts on both factors, $\mathcal{A}^*$ and $\mathbf{g}_E$, the quotient

$$\mathbf{P} = \mathcal{A}^* \times_{\widehat{\mathcal{G}}} \mathbf{g}_E \tag{2.52}$$

is a $G/C(G)$-bundle over $\mathcal{B}^* \times X$. This is the *universal bundle* associated with E (or V). In the case of $G = SU(2)$ or $SO(3)$ the universal bundle is an $SO(3)$ bundle (since $SU(2)/\mathbf{Z}_2 = SO(3)$ and $SO(3)$ has no center). Its Pontrjagin class $p_1(\mathbf{P})$ can be computed using Chern–Weil theory in terms of the curvature of a connection on $\mathbf{P}$. One can construct a natural connection on $\mathbf{P}$, called the *universal connection*, by considering the quotient of the tautological connection. The curvature of the universal connection will be denoted by $K_{\mathbf{P}}$. It is a form in $\Omega^2(\mathcal{B}^* \times X, \mathbf{g}_P)$, and splits according to the bi-grading of $\Omega^*(\mathcal{B}^* \times X)$ into three pieces: a two-form with respect to $\mathcal{B}^*$, a two-form with respect to X, and a mixed form (one-form on $\mathcal{B}^*$ and one-form on X), all with values in $\mathbf{g}_P$. The Pontrjagin class is:

$$\frac{p_1(\mathbf{P})}{4} = \frac{1}{8\pi^2}\mathrm{Tr}\ (K_{\mathbf{P}} \wedge K_{\mathbf{P}}). \tag{2.53}$$

By decomposing according to the bi-grading we obtain an element in $H^*(\mathcal{B}^*)\otimes H^*(X)$. To obtain differential forms on $\mathcal{B}^*$ we just take the slant product with *homology* classes in X (*i.e.*, we simply pair the forms on X with cycles on X). In this way we obtain the *Donaldson map*:

$$\mu : H_i(X) \longrightarrow H^{4-i}(\mathcal{B}^*). \tag{2.54}$$

One can prove that the differential forms obtained in this way actually generate the cohomology ring of $\mathcal{B}^*$. Finally, after restriction to $\mathcal{M}_{\mathrm{ASD}}$ we obtain the following differential forms on the moduli space of ASD connections:

$$\begin{aligned} x \in H_0(X) &\to \mathcal{O}(x) \in H^4(\mathcal{M}_{\mathrm{ASD}}),\\ \delta \in H_1(X) &\to I_1(\delta) \in H^3(\mathcal{M}_{\mathrm{ASD}}),\\ S \in H_2(X) &\to I_2(S) \in H^2(\mathcal{M}_{\mathrm{ASD}}). \end{aligned} \tag{2.55}$$

There are also cohomology classes associated with three-cycles in X, but we will not consider them in this book. In the next chapter we will see that the Donaldson map arises very naturally in the context of topological quantum field theory in what is called the *descent procedure*. In any case, we can now formally define the Donaldson invariants as follows. Consider the space

$$\mathbf{A}(X) = \mathrm{Sym}\ (H_0(X) \oplus H_2(X)) \otimes \wedge^* H_1(X), \tag{2.56}$$

with a typical element written as $x^\ell S_{i_1} \cdots S_{i_p}\delta_{j_1} \cdots \delta_{j_q}$. We have to choose an integer lifting w of the second Stiefel–Whitney class $w_2(V)$ of the gauge bundle. The *Donaldson invariant* corresponding to this element of $\mathbf{A}(X)$ is the following intersection number:

$$\mathcal{D}_X^{w,k}(x^\ell S_{i_1} \cdots S_{i_p}\delta_{j_1} \cdots \delta_{j_q}) =$$
$$\int_{\mathcal{M}_{\mathrm{ASD}}(w,k)} \mathcal{O}^\ell \wedge I_2(S_{i_1}) \wedge \cdots \wedge I_2(S_{i_p}) \wedge I_1(\delta_{j_1}) \wedge \cdots \wedge I_1(\delta_{j_q}),$$
$$\tag{2.57}$$

where we denoted by $\mathcal{M}_{\mathrm{ASD}}(w,k)$ the moduli space of ASD connections with second Stiefel–Whitney class $w_2(V)$ with lifting $w \in H^2(X, \mathbf{Z})$ and instanton number k. The choice of lifting w of $w_2(V)$ leads to a choice of orientation of the moduli space, and therefore the value of the invariant will depend on w, and not only on $w_2(V)$. However, this is a rather mild dependence: if $w, w' \in H^2(X, \mathbf{Z})$ are different choices of integer lifting then the corresponding Donaldson invariants differ by a sign given by

$$(-1)^{(w-w')\cdot w_2(X)/2}. \tag{2.58}$$

Notice that since the integrals of differential forms are different from zero only when the dimension of the space equals the total degree of the form it is clear that the integral in (2.57) will be different from zero only if the degrees of the forms add up to $\dim(\mathcal{M}_{\mathrm{ASD}}(w,k))$. It follows from (2.57) that Donaldson invariants can be understood as functionals:

$$\mathcal{D}_X^{w,k} : \mathbf{A}(X) \to \mathbf{Q}. \tag{2.59}$$

The reason that the values of the invariants are rational rather than integer is subtle and has to do with their being rigorously defined as intersection numbers only in certain situations (the so called *stable range*). Outside this range there is a natural way in which to extend the definition which involves dividing by 2.

It is very convenient to pack all Donaldson invariants in a generating functional. Let $\{\delta_i\}_{i=1,\ldots,b_1}$ be a basis of one-cycles, and $\{S_i\}_{i=1,\ldots,b_2}$ a basis of two-cycles. We introduce the formal sums

$$\delta = \sum_{i=1}^{b_1} \zeta_i\, \delta_i, \qquad S = \sum_{i=1}^{b_2} v_i\, S_i, \tag{2.60}$$

where v_i are complex numbers, and ζ_i are Grassmann variables. We then define the *Donaldson–Witten generating functional* as:

$$Z_{\mathrm{DW}}^w(p, \zeta_i, v_i) = \sum_{k=0}^{\infty} \mathcal{D}_X^{w,k}\big(\mathrm{e}^{px+\delta+S}\big), \tag{2.61}$$

where in the right hand side we are summing over all instanton numbers, *i.e.*, we are summing over all topological configurations of the $SO(3)$ gauge field with a *fixed* $w_2(V)$. This gives a formal power series in p, ζ_i, and v_i. The

Donaldson invariants can be extracted from the coefficients of these formal series. If we assign degree 4 to p, 2 to v_i, and 3 to ζ_i, and we fix the total degree, we obtain a finite polynomial which encodes all the Donaldson invariants for a fixed instanton number. Therefore Donaldson invariants at fixed instanton number can be also regarded as polynomials in the (dual of the) cohomology of the manifold. Sometimes we will also write (2.61) as a functional $Z_{\mathrm{DW}}^w(p, S, \delta)$.

The basic goal of Donaldson theory is the computation of the generating functional (2.61), and many results have been obtained over the years for different four-manifolds. More importantly, general structure results about the Donaldson–Witten generating functional are known, starting with the seminal work of Kronheimer and Mrowka. In order to explain these structure results we have to introduce some definitions. Let $w \in H^2(X, \mathbf{Z})$ be an integer lifting of the second Stiefel–Whitney class of V. A four-manifold X is said to be of *w-finite type* if there is an $n \geq 0$ such that

$$\left(\frac{\partial^2}{\partial p^2} - 4 \right)^n Z_{\mathrm{DW}}^w(p, S, \delta) = 0. \tag{2.62}$$

The order of w-finite type is the minimum of such n. It has been proved that if X is a four-manifold with $b_2^+ > 1$ and w, w' are elements in $H^2(X, \mathbf{Z})$, then the order of w-finite type and the order of w'-finite type are equal. Therefore we can just talk about manifolds of finite type and of their order without having to specify the choice of Stiefel–Whitney class. Moreover, it is known that *all* four-manifolds with $b_2^+ > 1$ are of finite type. A manifold is said to be of *Donaldson simple type* when it is of finite type with $n = 1$, *i.e.*, its Donaldson–Witten generating functional satisfies

$$\left(\frac{\partial^2}{\partial p^2} - 4 \right) Z_{\mathrm{DW}}^w(p, S, \delta) = 0. \tag{2.63}$$

From what we have said, this condition does not depend on the choice of w: if it is true for a particular choice it will be true for any other. Four-manifolds of simple type play a very special role in Donaldson theory owing to the following property. Define the *Donaldson series* as

$$\mathbf{D}^w(S) = Z_{\mathrm{DW}}^w(p, S)\big|_{p=0} + \frac{1}{2} \frac{\partial}{\partial p} Z_{\mathrm{DW}}^w(p, S)\big|_{p=0}, \tag{2.64}$$

where we have put $\delta = 0$. Notice that the Donaldson series can be regarded as a map

$$\mathbf{D}^w : \mathrm{Sym}_*(H_2(X)) \to \mathbf{Q}. \tag{2.65}$$

The most important structure result in Donaldson theory is that if a simply connected four-manifold X with $b_2^+ > 1$ is of simple type then the Donaldson series has the following structure:

$$\mathbf{D}^w(S) = \exp(S^2/2) \sum_{s=1}^{r} (-1)^{(w^2+\kappa_s\cdot w)/2} a_s e^{(\kappa_s,S)}, \qquad (2.66)$$

where the sum is over finitely many homology classes $\kappa_1, \ldots, \kappa_r \in H_2(X, \mathbf{Z})$ and non-zero rational numbers $a_1, \ldots, a_r$ which do not depend on w. Furthermore, each of the classes κ_i is characteristic. The classes κ_i are called *Donaldson basic classes*. This result shows, in particular, that the dependence on w is completely captured by the phase $(-1)^{(w^2+\kappa_s\cdot w)/2}$, therefore we will sometimes write the Donaldson series for the manifold X and $w = 0$ as a map

$$\mathbf{D}_X = e^{Q/2} \sum_{s=1}^{r} a_s e^{\kappa_s}, \qquad (2.67)$$

where Q is the intersection form and κ_s acts on an arbitrary two-homology class by intersection. Notice that the w dependence in (2.66) implies immediately that a change of lifting of $w_2(V)$ leads to the sign difference given by (2.58) (since κ_i are characteristic).

Many simply connected four-manifolds of $b_2^+ > 1$ are known to be of Donaldson simple type, and their Donaldson series have been computed. For example, the elliptic fibrations $E(n)$ with $n \geq 2$ are of Donaldson simple type. If we denote by f the homology class of the fiber (which is topologically a torus) one finds:

$$\mathbf{D}_{E(n)} = e^{Q/2} \sinh^{n-2}(f). \qquad (2.68)$$

This means in particular that the Donaldson basic classes are given by

$$\kappa_s = (n - 2s)f, \quad s = 1, \ldots, n - 1, \qquad (2.69)$$

and the coefficients a_s in the Donaldson series (2.67) are

$$a_s = (-1)^{s+1} 2^{2-n} \binom{n-2}{s-1}, \quad s = 1, \ldots, n - 1. \qquad (2.70)$$

Notice that $K3$ corresponds to $n = 2$, whose only Donaldson basic class is the trivial one.

In the non-simply connected case it is useful to introduce the condition of *strong simple type*. A four-manifold X is said to be of w-strong simple type if it is of Donaldson simple type and, moreover, $Z_{\mathrm{DW}}^w(p, S, \delta) = 0$ for $\delta \neq 0$. We will say that X is of strong simple type if it is of w-strong simple type for any $w \in H^2(X, \mathbf{Z})$. Again, if X is of w-strong simple type for some w, it is of strong simple type for all w, and its Donaldson series is again of the form (2.66). For example, the manifold $\Sigma_h \times \Sigma_g$ with $h, g \geq 1$ is of strong simple type. When $h = 1$ the complex manifold $Y_g = \mathbf{T}^2 \times \Sigma_g$ is an elliptic fibration, and its Donaldson series is given by

$$\mathbf{D}_{Y_g} = e^{Q/2} 4^g \sinh^{2g-2}(f), \tag{2.71}$$

where $f = [\mathbf{T}^2]$ is the class of the fiber. When $h, g \geq 2$, $Y_{g,h} = \Sigma_g \times \Sigma_h$ is an algebraic surface of general type, and its Donaldson series is given by

$$\mathbf{D}_{Y_{g,h}} = 2^{3+7(g-1)(h-1)} \begin{cases} \sinh K & g \text{ and } h \text{ even,} \\ \cosh K, & g \text{ or } h \text{ odd,} \end{cases} \tag{2.72}$$

where K is the class corresponding to the canonical line bundle of $Y_{g,h}$.

It is known that there are manifolds with $b_2^+ > 1$ which are not of strong simple type, but there are no known simply connected manifolds of $b_2^+ > 1$ which are not of Donaldson simple type. The conjecture that *all simply connected manifolds of $b_2^+ > 1$ are of Donaldson simple type* is one of the major open problems in the field. We refer the reader to the bibliography at the end of this chapter for more examples of manifolds of Donaldson simple type and computations of their Donaldson series.

One of the basic goals of Donaldson theory is to give an expression for the Donaldson–Witten generating function (2.61) as explicit as possible. As will be shown in this book, it is possible to derive the structure theorem (2.66) from topological quantum field theory, and, moreover, one can determine κ_i, a_i very precisely: as Witten discovered, these data involve some simpler invariants of four-manifolds, the Seiberg–Witten invariants, which will be the subject of the next chapter.

2.7. Metric dependence

Donaldson invariants turn out to be independent of the metric when $b_2^+ > 1$. This is closely related to the existence of reducible connections. Remember that a reducible connection gives a splitting of the $SU(2)$ bundle according to (2.33). In order for the reducible connection to solve the ASD equations, we have to solve the equation for an abelian instanton $F^+ = 0$.

This means, in particular, that $c_1(L)$ can be represented by an ASD harmonic two-form, and in particular it belongs to the intersection

$$H^{2,-}(X,\mathbf{R}) \cap H^2(X,\mathbf{Z}). \qquad (2.73)$$

For a generic metric, however, this intersection it just the zero element (unless $b_2^+ = 0$), since it is the intersection of an integer lattice in $H^2(X,\mathbf{R})$ with a proper subspace. Therefore for a generic metric there are no reducible ASD connections if $b_2^+ > 0$.

In order to test metric dependence, however, we have to see what happens to the moduli space of ASD connections as we move along a generic one-dimensional family of metrics (more precisely, along a one-dimensional family of conformal classes of metrics, since the instanton equation is conformally invariant). If $b_2^+ = 1$ then on a generic one-dimensional family we can in fact find a reducible ASD connection. This will provoke a singularity in the moduli space, and as consequence the Donaldson invariants will 'jump'. If, on the contrary, $b_2^+ > 1$, then we do not find reducible connections along generic one-dimensional families, and the Donaldson invariants will be truly metric-independent.

For $b_2^+ = 1$ the metric dependence can be described in more detail as follows. Let X be a four-manifold of $b_2^+ = 1$. The dependence on the metric is through the so called *period point* ω. The period point is defined as the harmonic two-form satisfying

$$*\omega = \omega, \qquad \omega^2 = 1. \qquad (2.74)$$

Clearly ω depends on the metric through the Hodge dual $*$. More precisely, it only depends on the conformal class of the metric (*i.e.*, rescaled metrics $g \to tg$ give the same ω). When we vary the conformal class of the metric we vary at the same time the period point ω in the space $H^2(X,\mathbf{R})$, and ω will describe a curve in the cone

$$V_+ = \{\omega \in H^2(X,\mathbf{R}) : \omega^2 > 0\}. \qquad (2.75)$$

This cone can be described in a fairly concrete way. For example, if we take $X_g = \mathbf{S}^2 \times \Sigma_g$ the general period point can be written as

$$\omega = \frac{1}{\sqrt{2}}(e^\theta[\mathbf{S}^2] + e^{-\theta}[\Sigma_g]), \qquad (2.76)$$

and ω describes a hyperbola in $H^2(X, \mathbf{R})$ parametrized by $-\infty < \theta < \infty$. Each point on this hyperbola corresponds to a choice of Kähler metric in $\mathbf{S}^2 \times \Sigma_g$. The limits $\theta \to \pm\infty$ correspond to limiting metrics which give a very small volume to $\mathbf{S}^2$ and Σ_g, respectively.

Imagine now that on a manifold of $b_2^+ = 1$ we start varying the period point ω in such a way that at a certain value there exists a cohomology class $\zeta \in H^2(X)$ which satisfies

$$\zeta \equiv w_2(V) \bmod 2, \quad \zeta^2 < 0, \quad (\zeta, \omega) = 0. \tag{2.77}$$

We then say that the element ζ defines a *wall* in V_+:

$$W_\zeta = \{\omega : (\zeta, \omega) = 0\}. \tag{2.78}$$

The complements of these walls are called *chambers*, and the cone V_+ is then divided in chambers separated by walls.

What is the meaning of these walls? If $\zeta \in H^2(X, \mathbf{Z})$ satisfies (2.77) then it is the first Chern class of a line bundle T which admits an ASD connection, since $\zeta^+ = (\zeta, \omega)\omega = 0$. The condition that ζ is congruent to $w_2(V)$ mod 2 is precisely the condition (2.37). Therefore ζ is the first Chern class associated with a reducible solution of the ASD equations, and it causes a singularity in moduli space: the Donaldson invariants jump when we pass through such a wall.

In summary, when $b_2^+ = 1$ the Donaldson invariants depend on the metric because they jump at walls, but they are metric independent *in each chamber*. We will represent by $Z_\pm^\zeta(p, S, \delta)$ the Donaldson–Witten generating function after and before passing the wall defined by ζ, respectively. One of the basic problems in Donaldson theory for manifolds of $b_2^+ = 1$ is to determine the jump in the generating function,

$$Z_+^\zeta(p, S, \delta) - Z_-^\zeta(p, S, \delta) = \mathrm{WC}_\zeta(p, S, \delta), \tag{2.79}$$

which is usually called the *wall crossing* term. It was conjectured by Kotschick and Morgan that the wall crossing term only depends on the classical homology ring of the four-manifold. Assuming the validity of this conjecture Göttsche was able to find a universal formula for $\mathrm{WC}(\zeta)$ in the simply connected case. As we will describe in Chapter 8, the u-plane integral allows an explicit evaluation of the wall crossing term that confirms the Kotschick-Morgan conjecture and reproduces Göttsche's formula. In the non-simply connected case only partial results are known mathematically. It will be also shown in Chapter 8 how by using the u-plane integral one can find a universal formula in the non-simply connected case as well.

Bibliographical notes

- The most complete survey of the theory of Donaldson invariants is the excellent book by Donaldson and Kronheimer [2]. Other useful resources include [6][7][8]. A nice and useful physicist's account can be found in [9]. Many of the results of Donaldson theory stated in this chapter are proved in detail in these references, which also provide a comprehensive bibliography.

- Torons were first considered by 't Hooft in [10].

- The Atiyah, Hitchin and Singer model for $\mathcal{M}_{\text{ASD}}$, together with the instanton deformation complex, is presented in [11].

- More details of the finite and simple type of conditions of the Donaldson series and of the structure theorem of Kronheimer and Mrowka can be found in [12] and [13], where the formula for the Donaldson series of the elliptic fibration $E(n)$ is also proved. A very nice review of computation of Donaldson invariants can be found in [14]. The finite type and strong simple type of conditions are discussed in [15], which also obtains the formula for the Donaldson series of $\Sigma_g \times \Sigma_h$ with $g, h > 1$ and proves that all four-manifolds with $b_2^+ > 1$ are of finite type. Donaldson invariants for manifolds which are not of simple type are explored in [16], [17] and [18]

- The conjecture by Kotschick and Morgan on the wall crossing term was presented in [19]. The universal formula for $\text{WC}(\zeta)$ in the simply connected case was found by Göttsche in [20]. Some partial results in the non-simply connected case are obtained in [21].

Chapter 3

The Theory of Seiberg–Witten Invariants

One of the major breakthroughs in the study of four-manifolds came in 1994, when Witten showed that the mathematical problem that motivated the introduction of Donaldson invariants —the classification of four-manifolds up to diffeomorphism— can be studied successfully by looking at a much simpler set of equations, the so called Seiberg–Witten equations. Witten also showed, basing on physical arguments, that Seiberg–Witten invariants contain all the information of Donaldson invariants, and that they provide the missing ingredient in the structure theorem of Kronheimer and Mrowka. In this section we will give a brief review of Seiberg–Witten invariants and their basic properties.

3.1. The Seiberg–Witten equations

Let X be an oriented closed Riemannian four-manifold, and let L be the determinant line bundle of a Spin_c structure. Positive chirality spinors, *i.e.*, sections of $S^+ \otimes L^{1/2}$, will be denoted by M. The Seiberg–Witten equations (also known as Seiberg–Witten monopole equations) are moduli equations for a pair (A, M) given by a $U(1)$ connection on L and a spinor M. Moduli equations for pairs consisting of a gauge connection and a section of some bundle have been considered many times (such as for example the Hitchin equations and vortex equations on Riemann surfaces). Although the rationale for introducing the Seiberg–Witten equations for pairs (A, M) comes from the analysis of a twisted $\mathcal{N} = 2$ supersymmetric theory, one can just write down the equations and explore their properties and the invariants which they define. This is what we are going to do in this chapter.

A section of the spinor bundle $M_{\dot{\alpha}}$ transforms in the **2** of $SU(2)_+$. On the other hand, F^+ is a SD form and it transforms in the **3**. In order to couple them we form a symmetric tensor in the **3** out of M_α by considering

$$\overline{M}_{(\dot\alpha}M_{\dot\beta)} = \begin{pmatrix} -M_1 M_2^* & \frac{1}{2}(|M_1|^2 - |M_2|^2) \\ \frac{1}{2}(|M_1|^2 - |M_2|^2) & M_1^* M_2 \end{pmatrix}. \tag{3.1}$$

The Seiberg–Witten equations are:

$$\begin{aligned} F_{\dot\alpha\dot\beta}^+ + 4i\overline{M}_{(\dot\alpha}M_{\dot\beta)} &= 0, \\ D_L^{\alpha\dot\alpha} M_{\dot\alpha} &= 0, \end{aligned} \tag{3.2}$$

where $F_{\dot\alpha\dot\beta}^+ = \overline{\sigma}_{\dot\alpha\dot\beta}^{\mu\nu} F_{\mu\nu}^+$, F is the curvature of the $U(1)$ connection on L, and D_L is the Dirac operator for the bundle $S^+ \otimes L^{1/2}$. Using (3.1) and the explicit form of $\overline{\sigma}_{\dot\alpha\dot\beta}^{\mu\nu}$ given in Appendix A, we can write the first equation as:

$$\begin{aligned} \frac{1}{2}(F_{12} + F_{34}) &= |M_1|^2 - |M_2|^2, \\ \frac{1}{2}(F_{13} + F_{42}) &= i(M_1 M_2^* - M_1^* M_2), \\ \frac{1}{2}(F_{14} + F_{23}) &= M_1 M_2^* + M_1^* M_2. \end{aligned} \tag{3.3}$$

3.2. The Seiberg–Witten invariants

The procedure for defining the Seiberg–Witten invariants is very similar to the procedure followed in Donaldson theory. First, one has to construct the moduli space of solutions of the equations (3.2). We will just present a local analysis, as we did in the case of $\mathcal{M}_{\mathrm{ASD}}$.

In the case of the Seiberg–Witten equations the configuration space is $\mathcal{C} = \mathcal{A} \times \Gamma(X, S^+ \otimes L^{1/2})$, where $\mathcal{A}$ is the moduli space of $U(1)$-connections on L. The group $\mathcal{G}$ of gauge transformations of the bundle L acts on this configuration space according to (2.4):

$$\begin{aligned} g^*(A_\mu) &= A_\mu + i\partial_\mu \log g, \\ g^*(M) &= gM, \end{aligned} \tag{3.4}$$

where $M \in \Gamma(X, S^+ \otimes L^{1/2})$ and g takes values in $U(1)$. The infinitesimal form of these transformations becomes, after putting $g = \exp(i\phi)$:

$$\begin{aligned} \delta A &= -d\phi, \\ \delta M &= i\phi M. \end{aligned} \tag{3.5}$$

The moduli space of solutions of the Seiberg–Witten equations, modulo gauge transformations, will be denoted by $\mathcal{M}_{\text{SW}}$. The tangent space of the configuration space at the point (A, M) is just $T_{(A,M)}\mathcal{C} = T_A\mathcal{A} \oplus T_M\Gamma(X, S^+ \otimes L^{1/2}) = \Omega^1(X) \oplus \Gamma(X, S^+ \otimes L^{1/2})$, since $\Gamma(X, S^+ \otimes L^{1/2})$ is a vector space.

A first step towards understanding the structure of the moduli space of solutions of the Seiberg–Witten equations modulo gauge transformations is to construct a slice of the gauge action, as we did in Donaldson theory. For this we need an explicit construction of the gauge orbits, given by the map $\mathcal{G} \times \mathcal{C} \to \mathcal{C}$. The space tangent to these orbits is the map

$$C : \text{Lie}\,(\mathcal{G}) \longrightarrow T\mathcal{C}. \qquad (3.6)$$

whose explicit expression in local coordinates can be obtained from (3.5):

$$C(\phi) = (-d\phi, i\phi M) \in \Omega^1(X) \oplus \Gamma(X, S^+ \otimes L^{1/2}), \quad \phi \in \Omega^0(X). \qquad (3.7)$$

The local model of the moduli space is then given by the zero locus in $\ker C^\dagger$ of the following map $s : \mathcal{C} \to \mathcal{F}$, where $\mathcal{F} = \Omega^{2,+}(X) \oplus \Gamma(X, S^- \otimes L^{1/2})$:

$$s(A, M) = \left(F^+_{\dot\alpha\dot\beta} + 4i\overline{M}_{(\dot\alpha}M_{\dot\beta)}, D^{\alpha\dot\alpha}_L M_{\dot\alpha} \right). \qquad (3.8)$$

As in Donaldson theory, we study the linearization of this map, $ds : T_{(A,M)}\mathcal{C} \to \mathcal{F}$. The explicit expression is:

$$ds(\psi, \mu) = \left((p^+(d\psi))_{\dot\alpha\dot\beta} + 4i\left(\overline{M}_{(\dot\alpha}\mu_{\dot\beta)} + \overline{\mu}_{(\dot\alpha}M_{\dot\beta)}\right), D^{\alpha\dot\alpha}_L\mu_{\dot\alpha} + \frac{i}{2}\psi^{\alpha\dot\alpha}M_{\dot\alpha} \right), \qquad (3.9)$$

where p^+ is the projector on SD forms, and we have written the connection and spinor as $A + \psi$, $M + \mu$. Instead of studying the restriction of this map to $\ker C^\dagger$ we can consider the instanton deformation complex:

$$0 \to \Omega^0(X) \xrightarrow{C} \Omega^1(X) \oplus \Gamma(X, S^+ \otimes L^{1/2}) \xrightarrow{ds} \Omega^{2,+}(X) \oplus \Gamma(X, S^- \otimes L^{1/2}) \to 0. \qquad (3.10)$$

which encodes the local model of the Seiberg–Witten moduli space and its virtual dimension. One can easily verify that (3.10) is, in fact, a complex, and after using the index theorem prove that the virtual dimension of Seiberg–Witten moduli space is

$$\dim \mathcal{M}_{\text{SW}} = \frac{1}{4}\left(c_1(L)^2 - 2\chi - 3\sigma\right). \qquad (3.11)$$

It is convenient to denote

$$\lambda = \frac{1}{2}c_1(L),\qquad(3.12)$$

which in general is not an integer class. Since $c_1(L) \equiv w_2(X) \bmod 2$ (it is characteristic), λ is an element in the lattice

$$H^2(X,\mathbf{Z}) + \frac{1}{2}w_2(X).\qquad(3.13)$$

From now on we will use λ to specify the topological class of the determinant line bundle involved in the Seiberg–Witten equations. In particular, the dimension of the moduli space (3.11) will be denoted by

$$d_\lambda = \lambda^2 - \frac{2\chi + 3\sigma}{4}.\qquad(3.14)$$

We then have a local model for the moduli space of solutions to the Seiberg–Witten moduli equations, which has virtual dimension (3.11). A more detailed analysis shows that this moduli space, in contrast to the ASD moduli space, is *compact*, and this makes life much easier!

The second step in defining the invariants is to construct a universal bundle, as in Donaldson theory. The procedure is very similar. Let $P_{U(1)}$ be the principal $U(1)$ bundle associated with the determinant line bundle $L \to X$ of a Spin$_c$-structure on X. We will denote by $\mathcal{M}^* \subset \mathcal{M} = \mathcal{A} \times \Gamma(X, S^+ \otimes L^{1/2})$ the subspace of irreducible pairs in the configuration space. On the space

$$\mathcal{M}^* \times P_{U(1)}\qquad(3.15)$$

there is an action of the gauge group $\mathcal{G} = \mathrm{Map}(X, U(1))$. The quotient

$$\mathbf{P}_{U(1)} = \mathcal{M}^* \times_{\mathcal{G}} P_{U(1)}\qquad(3.16)$$

is a $U(1)$-bundle over $(\mathcal{M}^*/\mathcal{G}) \times X$. This is the universal bundle associated with $P_{U(1)}$ for the Seiberg–Witten moduli problem. The first Chern class of this bundle is a closed two-form on the base $(\mathcal{M}^*/\mathcal{G}) \times X$,

$$c_1(\mathbf{P}_{U(1)}) = \frac{1}{2\pi}\mathcal{F},\qquad(3.17)$$

and therefore gives an element in $H^*(\mathcal{M}^*/\mathcal{G}) \otimes H^*(X)$. The analog of the Donaldson map is now

$$\mu : H_i(X) \longrightarrow H^{2-i}(\mathcal{M}^*/\mathcal{G}).\qquad(3.18)$$

The image of a point in X gives, after restriction to $\mathcal{M}_{\mathrm{SW}}$,

$$\phi \in H^2(\mathcal{M}_{\mathrm{SW}}). \tag{3.19}$$

Given a basis of one-cycles $\delta_1, \ldots, \delta_r \in H_1(X, \mathbf{Z})$ with duals $\beta_1, \ldots, \beta_r \in H^1(X, \mathbf{Z})$, we define the following one-forms:

$$\nu_i = \mu(\delta_i), \quad i = 1, \ldots, b_1, \tag{3.20}$$

again restricted to $\mathcal{M}_{\mathrm{SW}}$. We can now define the Seiberg–Witten invariant associated to a Spin_c-structure specified by $\lambda \in H^2(X, \mathbf{Z}) + \frac{1}{2}w_2(X)$ as a map

$$\mathrm{SW}_\lambda : \wedge^* H^1(X, \mathbf{Z}) \to \mathbf{Z} \tag{3.21}$$

defined by

$$\mathrm{SW}_\lambda(\beta_{i_1} \wedge \cdots \wedge \beta_{i_r}) = \int_{\mathcal{M}_\lambda} \nu_{i_1} \wedge \cdots \wedge \nu_{i_r} \wedge \phi^{\frac{1}{2}(d_\lambda - r)}. \tag{3.22}$$

Clearly $d_\lambda - r$ has to be even, otherwise the invariant is zero. We will denote

$$\mathrm{SW}(\lambda) = \int_{\mathcal{M}_\lambda} \phi^{d_\lambda/2} \tag{3.23}$$

as the Seiberg–Witten invariant in the case $r = 0$. This is the only relevant invariant in the simply connected case, and we can also regard it as a map from the characteristic elements in $H^2(X, \mathbf{Z})$ to $\mathbf{Z}$. We will say that λ is a *Seiberg–Witten basic class* if the map SW_λ is not identically zero. Of course, if λ is a basic class then 2λ is characteristic. In the simply connected case, a Seiberg–Witten basic class is just a class λ such that $\mathrm{SW}(\lambda) \neq 0$. A fundamental result of the theory of Seiberg–Witten invariants is that the number of basic classes on a four-manifold is *finite*.

We will say that a manifold X with $b_2^+ > 1$ is of *Seiberg–Witten simple type* if all basic classes have $d_\lambda = 0$. This means, in particular, that $\mathrm{SW}_\lambda(\beta_1 \wedge \cdots \wedge \beta_r) = 0$ for $r > 0$, and the Seiberg–Witten invariants reduce to (3.23). All known simply connected manifolds with $b_2^+ > 1$ are of Seiberg–Witten simple type. It is easy to see by using (1.27) that if a manifold is of Seiberg–Witten simple type then

$$\chi_h = \frac{\chi + \sigma}{4} \tag{3.24}$$

is an integer number. There are many classes of manifolds for which χ_h is always an integer. For example, on complex manifolds one has:

$$\chi_h = 1 - h^{1,0} + h^{2,0}, \tag{3.25}$$

where $h^{p,q}$ is the dimension of $H^{p,q}(X)$, and χ_h is known as the holomorphic Euler characteristic. It is possible to show that if λ is a Seiberg–Witten basic class then $-\lambda$ is a Seiberg–Witten basic class as well, and, furthermore, in the simple type case their Seiberg–Witten invariants are related as follows:

$$\mathrm{SW}(-\lambda) = (-1)^{\chi_h}\mathrm{SW}(\lambda). \tag{3.26}$$

Another important property of Seiberg–Witten invariants is the following vanishing result: if X is a four-manifold of $b_2^+ > 0$ which admits a Riemannian metric of positive curvature, then the Seiberg–Witten invariants necessarily vanish. This follows from the Weitzenböck formula for the Dirac operator.

One of the basic advantages of Seiberg–Witten invariants is that they are relatively easy to compute. To end this section we will give some examples of Seiberg–Witten invariants of four-manifolds. The elliptic fibrations $E(n)$ with $n > 1$ turn out to be of Seiberg–Witten simple type, and the basic classes are

$$2\lambda_s = (n - 2s)f, \quad s = 1, \ldots, n - 1, \tag{3.27}$$

and their Seiberg–Witten invariants are given by

$$\mathrm{SW}(\lambda_s) = (-1)^{s+1}\binom{n-2}{s-1}, \quad s = 1, \ldots, n - 1. \tag{3.28}$$

Notice that the Seiberg–Witten basic classes agree with the Donaldson basic classes shown in (2.69).

Another class of four-manifolds of Seiberg–Witten simple type are minimal surfaces of general type with $b_2^+ > 1$. In that case the only Seiberg–Witten basic classes turn out to be $2\lambda = \pm K$, where K is the canonical line bundle, and their Seiberg–Witten invariants are given respectively $(-1)^{\chi_h}$ (with χ_h given by (3.24)) and 1.

3.3. Metric dependence

Like Donaldson invariants, Seiberg–Witten invariants exhibit metric dependence when $b_2^+ = 1$. The reason is also very similar. If we look at (3.4) we see that reducible pairs (*i.e.*, pairs (A, M) with a non-trivial isotropy group) must have $M = 0$. Therefore, reducible pairs that satisfy the Seiberg–Witten equations are just abelian instantons $F^+ = 0$. This means that if $b_2^+ = 1$, as we move along a one-dimensional family we will find a reducible solution of the Seiberg–Witten equations, and therefore the value of the Seiberg–Witten invariants will change. As in Donaldson theory, the metric dependence has a structure of chambers and walls in the cone V_+. However, the walls are defined in this case by the conditions:

$$2\lambda \equiv w_2(X) \,\mathrm{mod}\, 2, \quad \lambda^2 < 0, \quad (\lambda, \omega) = 0, \quad d_\lambda \geq 0, \qquad (3.29)$$

where 2λ is the first Chern class of the determinant line bundle of a Spin$_c$ structure. The third condition in (3.29) tells us that this line bundle provides a reducible solution of the Seiberg–Witten equations: $M = 0$, $F^+ = 0$, and this causes a jump in the value of the Seiberg–Witten invariant, so we have a wall crossing term:

$$\mathrm{SW}_\lambda^+(\beta_1 \wedge \cdots \beta_r) - \mathrm{SW}_\lambda^-(\beta_1 \wedge \cdots \beta_r) = \mathrm{WC}_\lambda(\beta_1 \wedge \cdots \beta_r). \qquad (3.30)$$

In this case, life is much simpler than in Donaldson theory. It is not very difficult to show that, when X is simply connected

$$\mathrm{SW}_+(\lambda) - \mathrm{SW}_-(\lambda) = \pm 1. \qquad (3.31)$$

The case in which X is not simply connected has also been worked out. We will discuss the general formula for wall crossing of Seiberg–Witten invariants when we discuss the u-plane integral.

Bibliographical notes

- Seiberg–Witten invariants were introduced by Witten in [22], motivated by the work of Seiberg and Witten [23][24] on $\mathcal{N} = 2$, $SU(2)$ Yang–Mills theory. A more detailed mathematical treatment can be found in [25][1][26][27]. A beautiful review of the implications of Seiberg–Witten invariants on four-manifold topology can be found in [28]. Seiberg–Witten invariants are reviewed from the physical point of view in [29][30][31]. The relation between Seiberg–Witten equations and vortex equations is discussed in [32].

- The extension of Seiberg–Witten invariants to the non-simply connected case (and in particular the definition of the general invariant defined in (3.22)) is considered, for example, in [33].

- The Seiberg–Witten invariants of algebraic surfaces are computed in [34], [35] and [36]. Seiberg–Witten invariants turn out to be particularly useful in the study of symplectic manifolds thanks to the work of Taubes [37], see [1], Chapter 9, for a detailed overview.

- The wall crossing formula for Seiberg–Witten invariants in the simply connected case is obtained in [22] and [38]. The general wall crossing formula in the non-simply connected case is obtained in [39] and [33].

Chapter 4

Supersymmetry in Four Dimensions

This chapter deals with supersymmetry in four dimensions. It contains a description of all the supersymmetric theories which will be used in future chapters. The scope of this description is to provide the elements of these theories which are needed for constructing topological quantum field theories. After a brief introduction to the supersymmetry algebra and to the notions of superspace and superfield, all the relevant models for the scope of this book involving supersymmetric Yang–Mills theory in four dimensions are considered and formulated. In this chapter we will work in Minkowski space with the metric $\eta_{\mu\nu} = \text{diag}(-1, 1, 1, 1)$. A summary of our conventions is contained in Appendix A.

4.1. The supersymmetry algebra

Supersymmetry is the only non-trivial extension of Poincaré symmetry that is compatible with the general principles of relativistic quantum field theory. Besides the ordinary generators of the Poincaré group the supersymmetric algebra possesses $\mathcal{N}$ fermionic generators,

$$Q_{\alpha u}, \qquad \overline{Q}^u_{\dot\alpha}, \qquad u = 1, \ldots, \mathcal{N}, \tag{4.1}$$

which transform in the spinorial representations S and $\overline{S}$, respectively, under the Lorentz group (see Appendix A for conventions regarding spinors). The resulting super-Poincaré algebra extends the usual Poincaré algebra introducing anticommutators for these fermionic generators.

The part of the algebra involving the fermionic generators has the form:

$$\{Q_{\alpha u}, \overline{Q}^v_{\dot\beta}\} = 2\delta^v_u \sigma^\mu_{\alpha\dot\beta} P_\mu, \qquad \{Q_{\alpha u}, Q_{\beta v}\} = 2\sqrt{2}\epsilon_{\alpha\beta} Z_{uv},$$

$$[P_\mu, Q_{\alpha v}] = 0, \qquad [P_\mu, \overline{Q}^u_{\dot\alpha}] = 0, \tag{4.2}$$

$$[M^{\mu\nu}, Q_{\alpha u}] = -(\sigma^{\mu\nu})_\alpha{}^\beta Q_{\beta u}, \qquad [M^{\mu\nu}, \overline{Q}^{\dot\alpha u}] = -(\overline{\sigma}^{\mu\nu})^{\dot\alpha}{}_{\dot\beta} \overline{Q}^{\dot\beta u},$$

where $u, v = 1, \ldots, \mathcal{N}$. The quantities $\sigma^\mu_{\alpha\dot\alpha}$, $(\sigma^{\mu\nu})_\alpha{}^\beta$ and $(\bar\sigma^{\mu\nu})^{\dot\alpha}{}_{\dot\beta}$ which appear on the right hand side of these equations are defined in Appendix A. The terms Z_{uv} are the so called central charges. They satisfy

$$Z_{uv} = -Z_{vu}, \tag{4.3}$$

and they commute with all the generators of the algebra that appear in (4.2). Property (4.3) implies that $\mathcal{N} \geq 2$ in order to have central charges.

When the central charges vanish, the theory has an internal $U(\mathcal{N})_R$ symmetry:

$$Q_{\alpha v} \to U_v{}^w Q_{\alpha w}, \qquad \overline{Q}^v_{\dot\alpha} \to \overline{U}_w{}^v \overline{Q}^w_{\dot\alpha}, \tag{4.4}$$

where $U \in U(\mathcal{N})_R$, and $\overline{U}$ is the complex conjugate of U. The generators of this symmetry will be denoted by B_a. Their commutation relations with the fermionic generators are:

$$[Q_{\alpha v}, B^a] = (b^a)_v{}^w Q_{\alpha w}, \qquad [\overline{Q}^w_{\dot\alpha}, B^a] = -\overline{Q}^v_{\dot\alpha}(b^a)_v{}^w, \tag{4.5}$$

where $b^a = b^{a\dagger}$. The central charges are linear combinations of the $U(\mathcal{N})_R$ generators,

$$Z_{uv} = d_{uv}{}^a B^a. \tag{4.6}$$

If the central charges are not zero the internal symmetry becomes reduced to $\text{USp}(\mathcal{N})$, formed by the unitary transformations that leave invariant the 2-form (4.6) in $\mathcal{N}$ dimensions. The $U(1)_R$ of the internal symmetry (4.4), with generator R, leads to the following chiral symmetry,

$$[Q_{\alpha v}, R] = Q_{\alpha v}, \qquad [\overline{Q}^w_{\dot\alpha}, R] = -\overline{Q}^v_{\dot\alpha}. \tag{4.7}$$

This symmetry is typically anomalous quantum mechanically, and quantum effects break it down to a discrete subgroup. Explicit realizations of this symmetry will appear when considering supersymmetric Yang–Mills theories with extended supersymmetry.

4.2. $\mathcal{N} = 1$ superspace and superfields

In order to find a local realization of supersymmetry one has to extend the usual Minkowski space to the so called *superspace*. In this section we are going to develop the basics of $\mathcal{N} = 1$ superspace, a framework which is extremely useful for formulating supersymmetric field theories. The first task is to construct a local realization of the supersymmetry algebra (4.2) for the case in which there are only two fermionic generators Q_α, $\overline{Q}^{\dot\alpha}$.

Superspace is obtained by adding four spinor coordinates, $\theta^\alpha, \overline{\theta}_{\dot\alpha}$, to the four space-time coordinates x^μ. The generator of supersymmetric transformations in superspace is

$$i\xi^\alpha Q_\alpha + i\overline{\xi}_{\dot\alpha}\overline{Q}^{\dot\alpha} = i(\xi Q - \overline{\xi}\,\overline{Q}),\tag{4.8}$$

where, $\xi^\alpha, \overline{\xi}_{\dot\alpha}$, are (fermionic) transformation parameters, and $Q_\alpha, \overline{Q}_{\dot\alpha}$ furnish a representation of the algebra (4.2). A standard choice in terms of derivatives on superspace is provided by,

$$Q_\alpha = -i\left(\frac{\partial}{\partial\theta^\alpha} - i\sigma^\mu_{\alpha\dot\alpha}\overline{\theta}^{\dot\alpha}\,\partial_\mu\right),\qquad \overline{Q}_{\dot\alpha} = i\left(\frac{\partial}{\partial\overline{\theta}^{\dot\alpha}} - i\theta^\alpha\sigma^\mu_{\alpha\dot\alpha}\,\partial_\mu\right).\tag{4.9}$$

They satisfy the anticommutation relations,

$$\{Q_\alpha, \overline{Q}_{\dot\alpha}\} = -2i\sigma^\mu_{\alpha\dot\alpha}\,\partial_\mu,\tag{4.10}$$

and therefore since $P_\mu = -i\partial_\mu$ they generate a representation of the supersymmetry algebra (4.2) for $\mathcal{N} = 1$. Under these generators superspace coordinates transform as,

$$\begin{aligned} x^\mu \to x'^\mu &= x^\mu + i\theta\sigma^\mu\overline{\xi} - i\xi\sigma^\mu\overline{\theta}\,,\\ \theta \to \theta' &= \theta + \xi\,,\\ \overline{\theta} \to \overline{\theta}' &= \overline{\theta} + \overline{\xi}\,. \end{aligned}\tag{4.11}$$

A *superfield* is just a function $H(x,\theta,\overline{\theta})$ on superspace. Since the θ-coordinates are anti-commuting the Taylor expansion in the fermionic coordinates truncates after a finite number of terms. Therefore, the most general $\mathcal{N} = 1$ superfield can always be expanded as

$$\begin{aligned} H(x,\theta,\overline{\theta}) =&f(x) + \theta\phi(x) + \overline{\theta}\overline{\chi}(x) + \theta\theta m(x) + \overline{\theta}\overline{\theta} n(x) + \theta\sigma^\mu\overline{\theta} v_\mu(x)\\ &+ \theta\theta\overline{\theta}\overline{\lambda}(x) + \overline{\theta}\overline{\theta}\theta\psi(x) + \theta\theta\overline{\theta}\overline{\theta} d(x)\,. \end{aligned}\tag{4.12}$$

In this expression $f(x)$, $\phi(x)$, etc., are ordinary fields and are called component fields of the superfield H. They are grouped into multiplets which correspond to irreducible representations of supersymmetry. Under the supersymmetry transformations (4.8), superfields transform as

$$\delta H = i(\xi Q - \overline{\xi}\,\overline{Q})H,\tag{4.13}$$

and from this expression one can obtain the transformation of the component fields.

In order to define constraints on superfields it is useful to define super-covariant derivatives. A standard choice is the following:

$$D_\alpha = \frac{\partial}{\partial \theta^\alpha} + i\sigma^\mu_{\alpha\dot\alpha} \bar\theta^{\dot\alpha} \partial_\mu \,, \qquad \overline{D}_{\dot\alpha} = -\frac{\partial}{\partial \bar\theta^{\dot\alpha}} - i\sigma^\mu_{\alpha\dot\alpha}\theta^\alpha \, \partial_\mu \,, \qquad (4.14)$$

which satisfy

$$\{D_\alpha, \overline{D}_{\dot\alpha}\} = -2i\sigma^\mu_{\alpha\dot\alpha}\, \partial_\mu, \qquad \{D_\alpha, \overline{Q}_{\dot\alpha}\} = \{Q_\alpha, \overline{D}_{\dot\alpha}\} = 0. \qquad (4.15)$$

The generic superfield gives a reducible representation of the supersymmetry algebra. In order to obtain irreducible representations one must impose constraints. This can be done with the help of the super-covariant derivatives (4.14). There are two different $\mathcal{N} = 1$ irreducible multiplets: the chiral multiplet and the vector multiplet. We will consider the first one in this section. The other one leads to $\mathcal{N} = 1$ supersymmetric Yang–Mills theory, which is the subject of the next section.

The chiral multiplet is represented by a superfield Φ which satisfies the following constraint:

$$\overline{D}_{\dot\alpha}\Phi = 0. \qquad (4.16)$$

Superfields satisfying this constraint are also called chiral superfields. The constraint can be easily solved by noting that if $y^\mu = x^\mu + i\theta\sigma^\mu\bar\theta$, then

$$\overline{D}_{\dot\alpha} y^\mu = 0, \quad \overline{D}_{\dot\alpha}\theta^\beta = 0 \,. \qquad (4.17)$$

Therefore, any function of (y, θ) is a chiral superfield. We can then write

$$\Phi(y, \theta) = \phi(y) + \sqrt{2}\theta^\alpha \psi_\alpha(y) + \theta^2 F(y) \,, \qquad (4.18)$$

and conclude that a chiral superfield contains two complex scalar fields, ϕ and F, and a spinor ψ_α. These are called component fields. In a similar way we can define an anti-chiral superfield by $D_\alpha \Phi^\dagger = 0$, which can be expanded as

$$\Phi^\dagger(y^\dagger, \bar\theta) = \phi^\dagger(y^\dagger) + \sqrt{2}\bar\theta\bar\psi(y^\dagger) + \bar\theta^2 F^\dagger(y^\dagger) \,, \qquad (4.19)$$

where, $y^{\mu\dagger} = x^\mu - i\theta\sigma^\mu\bar\theta$. Again, its field content is a spinor and two complex scalars.

Supersymmetric transformations are computed using the generator (4.8), which acts on superfields in the manner stated in (4.13). Defining the supersymmetric transformations of the components fields by the relation:

$$\delta\Phi(y,\theta) = \delta\phi(y) + \sqrt{2}\theta^\alpha \delta\psi_\alpha(y) + \theta^2 \delta F(y) \qquad (4.20)$$

one finds

$$\delta\phi = \sqrt{2}\xi^\alpha \psi_\alpha,$$
$$\delta\psi_\alpha = \sqrt{2}\xi_\alpha F + i\sqrt{2}\overline{\xi}^{\dot\alpha}\sigma^\mu_{\alpha\dot\alpha}\partial_\mu \phi, \qquad (4.21)$$
$$\delta F = i\sqrt{2}\overline{\xi}^{\dot\alpha}\sigma^\mu_{\alpha\dot\alpha}\partial_\mu \psi^\alpha.$$

In terms of the original variables, the chiral superfields Φ and $\Phi^\dagger$ take the form

$$
\begin{aligned}
\Phi(x,\theta,\overline{\theta}) &= \phi(x) + i\theta\sigma^\mu\overline{\theta}\partial_\mu\phi - \frac{1}{4}\theta^2\overline{\theta}^2\nabla^2\phi \\
&\quad + \sqrt{2}\theta\psi(x) - \frac{i}{\sqrt{2}}\theta\theta\partial_\mu\psi\sigma^\mu\overline{\theta} + \theta\theta F(x)\,, \\
\Phi^\dagger(x,\theta,\overline{\theta}) &= \phi^\dagger(x) - i\theta\sigma^\mu\overline{\theta}\partial_\mu\phi^\dagger - \frac{1}{4}\theta^2\overline{\theta}^2\nabla^2\phi^\dagger \\
&\quad + \sqrt{2}\overline{\theta\psi}(x) + \frac{i}{\sqrt{2}}\overline{\theta}\overline{\theta}\,\theta\sigma^\mu\,\partial_\mu\overline{\psi} + \overline{\theta}\overline{\theta}F^\dagger(x)\,.
\end{aligned}
\qquad (4.22)
$$

Actions in superspace are constructed using the Berezin integral. For each anticommuting coordinate θ this integral is defined as a linear operation on superfields such that

$$\int d\theta = 0, \qquad \int d\theta\,\theta = 1. \qquad (4.23)$$

We use conventions such that for more than one θ one has $\int d^2\theta\,\theta^2 = 1$ and $\int d^2\overline{\theta}\,\overline{\theta}^2 = 1$. Notice that θ-integrals pick up the highest component of the superfield. The Berezin integral of a superfield is invariant under supersymmetric transformations when one integrates over the full measure $d^2\theta\,d^2\overline{\theta}$. One can also obtain invariant actions integrating over the chiral (anti-chiral) measure, $d^2\theta$ ($d^2\overline{\theta}$), when the integrand is a chiral (anti-chiral) superfield. The action for the kinetic part of the chiral multiplet takes the form:

$$\frac{1}{4}\int d^4x\,d^2\theta\,d^2\overline{\theta}\,\Phi^\dagger\Phi = \int d^4x\,(\phi^\dagger\nabla^2\phi - \frac{i}{2}\overline{\psi}^{\dot\alpha}\sigma^\mu_{\alpha\dot\alpha}\partial_\mu\psi^\alpha - F^\dagger F), \qquad (4.24)$$

which leads to the standard kinetic terms for a complex scalar ϕ and a spinor ψ_α. The field F turns out to be an auxiliary field.

Among the possible terms that could be added to the previous action it is worth to describe the one that leads to mass terms. It has the form:

$$\frac{m}{2}\int d^4x\, d^2\theta\,\Phi^2 + \frac{m}{2}\int d^4x\, d^2\overline{\theta}\,\overline{\Phi}^2 =$$
$$m\int d^4x\left(\phi F + \phi^\dagger F^\dagger + \frac{1}{2}\left(\psi^\alpha\psi_\alpha + \overline{\psi}_{\dot\alpha}\overline{\psi}^{\dot\alpha}\right)\right). \qquad (4.25)$$

Considering the sum of the actions (4.24) and (4.25), and integrating out the auxiliary field F, one obtains the standard action of a massive complex scalar field ϕ and a massive spinor ψ_α:

$$\int d^4x\left(\phi^\dagger\nabla^2\phi - \frac{i}{2}\overline{\psi}^{\dot\alpha}\sigma^\mu_{\alpha\dot\alpha}\partial_\mu\psi^\alpha - m\phi^\dagger\phi - \frac{m}{2}(\psi^\alpha\psi_\alpha + \overline{\psi}_{\dot\alpha}\overline{\psi}^{\dot\alpha})\right). \qquad (4.26)$$

Now the supersymmetry transformations (4.21) simplify to

$$\delta\phi = \sqrt{2}\xi^\alpha\psi_\alpha,$$
$$\delta\psi_\alpha = i\sqrt{2}\overline{\xi}^{\dot\alpha}\sigma^\mu_{\alpha\dot\alpha}\partial\phi, \qquad (4.27)$$

but they close only on-shell, *i.e.*, one needs to use the field equations to recover the supersymmetric algebra.

The superspace formalism allows to obtain very easily the most general action involving several chiral superfields. Let us denote by Φ a set of n chiral superfields, $\Phi = (\Phi_1, \ldots, \Phi_n)$. The most general $\mathcal{N} = 1$ supersymmetric Lagrangian for the scalar multiplet (including the interaction terms) is given by

$$\mathcal{L} = \int d^4\theta\, K(\Phi, \Phi^\dagger) + \int d^2\theta\, \mathcal{W}(\Phi) + \int d^2\overline{\theta}\,\overline{\mathcal{W}}(\Phi^\dagger), \qquad (4.28)$$

where K and $\mathcal{W}$ are arbitrary functions. The kinetic term for the scalar field ϕ_i (the θ- independent component of Φ_i) takes the form,

$$g^{ij}\partial_\mu\phi_i\partial^\mu\phi_j^\dagger, \qquad (4.29)$$

where

$$g^{ij} = \frac{\partial^2 K}{\partial\Phi_i\partial\Phi_j^\dagger}, \qquad (4.30)$$

is in general a non-trivial metric for the space of fields Φ. This has the form of a Kähler metric derived from a Kähler potential $K(\Phi_i, \Phi_j^\dagger)$. For this reason, the function $K(\Phi, \Phi^\dagger)$ is referred to as the *Kähler potential*. The simplest Kähler potential, corresponding to the flat metric, is

$$K(\Phi, \Phi^\dagger) = \sum_{i=1}^{n}\Phi_i^\dagger\Phi_i,$$

which leads to the free action (4.24) for a set of n massless scalars and n massless spinors.

The function $\mathcal{W}(\Phi)$ in (4.28) is an arbitrary holomorphic function of chiral superfields, and it is called the *superpotential*. It can be expanded as,

$$\mathcal{W}(\Phi_i) = \mathcal{W}(A_i + \sqrt{2}\theta\psi_i + \theta\theta F_i)$$
$$= \mathcal{W}(A_i) + \frac{\partial \mathcal{W}}{\partial A_i}\sqrt{2}\theta\psi_i + \theta\theta\left(\frac{\partial \mathcal{W}}{\partial A_i}F_i - \frac{1}{2}\frac{\partial^2 \mathcal{W}}{\partial A_i A_j}\psi_i\psi_j\right). \tag{4.31}$$

Supersymmetric interaction terms can be constructed in terms of the superpotential and its conjugate. Finally, we have to mention that there is $U(1)_R$ symmetry that acts as follows:

$$R\,\Phi(x,\theta) = e^{2in\alpha}\Phi(x, e^{-i\alpha}\theta)\,,$$
$$R\,\Phi^\dagger(x,\bar\theta) = e^{-2in\alpha}\Phi^\dagger(x, e^{i\alpha}\bar\theta)\,. \tag{4.32}$$

Under this the component fields transform as

$$A \to e^{2in\alpha}A\,,$$
$$\psi \to e^{2i(n-1/2)\alpha}\psi\,, \tag{4.33}$$
$$F \to e^{2i(n-1)\alpha}F\,.$$

4.3. $\mathcal{N}=1$ supersymmetric Yang–Mills theories

The other basic $\mathcal{N}=1$ multiplet is the vector multiplet. It is the main ingredient in $\mathcal{N}=1$ supersymmetric Yang–Mills theories. We will discuss first the abelian version of this multiplet. Then we will deal with the non-abelian case and its coupling to chiral multiplets.

The $\mathcal{N}=1$ vector multiplet is represented by a real superfield satisfying $V=V^\dagger$. In components, it has the following expansion:

$$V(x,\theta,\bar\theta) = C + i\theta\chi - i\bar\theta\bar\chi + \frac{1}{2}\theta^2(M+iN) + \frac{1}{2}\bar\theta^2(M-iN) - \theta\sigma^\mu\bar\theta A_\mu$$
$$+ i\theta^2\bar\theta\left(\bar\lambda + \frac{i}{2}\bar\sigma^\mu\partial_\mu\chi\right) - i\bar\theta^2\theta\left(\lambda + \frac{i}{2}\sigma^\mu\partial_\mu\bar\chi\right) + \frac{1}{2}\theta^2\bar\theta^2\left(D - \frac{1}{2}\nabla^2 C\right). \tag{4.34}$$

By performing an abelian gauge transformation

$$V \to V + i(\Lambda - \Lambda^\dagger)\,, \tag{4.35}$$

where Λ ($\Lambda^\dagger$) are chiral (anti-chiral) superfields, one can set $C = M = N = \chi = 0$. This is the so called Wess–Zumino gauge. In this gauge the superfield V takes the form,

$$V = -\theta\sigma^\mu\bar\theta A_\mu + i\theta^2\bar\theta\bar\lambda - i\bar\theta^2\theta\lambda + \frac{1}{2}\theta^2\bar\theta^2 D\,, \tag{4.36}$$

and it turns out that $V^2 = \frac{1}{2} A_\mu A^\mu \theta^2 \overline{\theta}^2$ and $V^3 = 0$. The Wess–Zumino gauge breaks all the component gauge symmetries in (4.35) except the one of the abelian gauge field A_μ. The abelian superfield strength is defined by,

$$W_\alpha = -\frac{1}{4}\overline{D}^2 D_\alpha V , \qquad \overline{W}_{\dot{\alpha}} = -\frac{1}{4}D^2 \overline{D}_{\dot{\alpha}} V. \tag{4.37}$$

It is gauge invariant, and W_α ($\overline{W}_{\dot{\alpha}}$) is a chiral (anti-chiral) spinorial superfield. In the Wess–Zumino gauge W_α takes the form

$$W_\alpha = -i\lambda_\alpha(y) + \theta_\alpha D - \frac{i}{2}(\sigma^\mu \overline{\sigma}^\nu \theta)_\alpha F_{\mu\nu} + \theta^2 (\sigma^\mu \partial_\mu \overline{\lambda})_\alpha . \tag{4.38}$$

In order to obtain the supersymmetric transformations of the component fields one needs to take into account restoring gauge transformations to remain in the Wess–Zumino gauge. This task is carried out considering the full (supersymmetric (4.13) + gauge (4.35)) transformation of the real superfield V,

$$\delta V = i(\xi Q - \overline{\xi}\overline{Q})V + i(\Lambda - \Lambda^\dagger), \tag{4.39}$$

and solving for the components of Λ imposed by $C = M = N = \chi = 0$. To carry this out one uses all the components of Λ (which has an expansion as the chiral superfield (4.18)) except the real part of its θ-independent component. Denoting this real part by $v/2$, the supersymmetric and gauge transformations of the component fields which remain in the Wess–Zumino gauge (see (4.36)) are:

$$\delta A^\mu = \partial^\mu v - i\sigma^\mu_{\alpha\dot{\alpha}}(\xi^\alpha \overline{\lambda}^{\dot{\alpha}} - \overline{\xi}^{\dot{\alpha}} \lambda^\alpha),$$

$$\delta \lambda_\alpha = i\xi_\alpha D - i\sigma^{\mu\nu\,\beta}_\alpha \xi_\beta F_{\mu\nu}, \tag{4.40}$$

$$\delta D = -\sigma^\mu_{\alpha\dot{\alpha}}(\xi^\alpha \partial_\mu \overline{\lambda}^{\dot{\alpha}} - \overline{\xi}^{\dot{\alpha}} \partial_\mu \lambda^\alpha),$$

where, $F_{\mu\nu} = \partial_\mu A_\nu - \partial_\nu A_\mu$, is the ordinary Yang–Mills field strength. These supersymmetric transformations close up to gauge transformations, as expected from the fact that one works in the Wess–Zumino gauge.

The action of the theory is obtained using the gauge invariant spinorial superfield W_α. Since it is chiral and one needs a scalar, the only quadratic choices are the real and imaginary parts of

$$\int d^4x \, d^2\theta \, W^\alpha W_\alpha . \tag{4.41}$$

A convenient way to parametrize the most general quadratic action is achieved introducing the complex parameter:

$$\tau = \frac{\Theta}{2\pi} + i\frac{4\pi}{g^2}, \tag{4.42}$$

where g and Θ are real parameters. We have used in (4.42) Θ instead of the usual parameter, θ, to avoid confusion with the superspace coordinates. The action in superspace takes the form

$$\frac{1}{4\pi}\mathrm{Im}\left(\tau\int d^4x\,d^2\theta\,W^\alpha W_\alpha\right). \tag{4.43}$$

This leads to the free action of $\mathcal{N}=1$ supersymmetric Yang–Mills theory in the abelian case:

$$\frac{1}{g^2}\int d^4x\left(-\frac{1}{4}F^{\mu\nu}F_{\mu\nu}-i\lambda^\alpha\sigma^\mu_{\alpha\dot\alpha}\partial_\mu\overline\lambda^{\dot\alpha}+\frac{1}{2}D^2\right)-\frac{\Theta}{64\pi^2}\epsilon^{\mu\nu\rho\sigma}F_{\mu\nu}F_{\rho\sigma}, \tag{4.44}$$

where $\epsilon^{\mu\nu\rho\sigma}$ is the totally antisymmetric tensor associated to the volume form with $\epsilon^{1234}=1$.

We will now study the non-abelian case. Let G be a gauge group and let us consider a set of real vector superfields V^a, $a=1,\ldots,\dim G$. If T^a, $a=1,\ldots,\dim G$, are group generators one constructs the group-valued superfield $V=V^aT^a$. The non-abelian generalization of the gauge transformation (4.35) is:

$$e^{2V}\to e^{-i\Lambda^\dagger}e^{2V}e^{i\Lambda}, \tag{4.45}$$

where $\Lambda=\Lambda^aT^a$, and Λ^a, $a=1,\ldots,\dim G$, are chiral superfields. The factor 2 in e^{2V} is introduced for convenience. With this choice covariant derivatives with the standard normalizations will appear naturally. Notice that to first order the transformation (4.45) matches (4.35). This transformation can be also used to set the components $C^a=M^a=N^a=\chi^a=0$, $a=1,\ldots,\dim G$, and thus one can also work in this case in the Wess–Zumino gauge. In this gauge the superfields V^a acquire the same form as their abelian counterpart in (4.36), and $V^n=0$ for $n>2$. Thus e^{2V}, the basic object of the theory, has the simple expansion,

$$e^{2V}=1+2V+2V^2. \tag{4.46}$$

The non-abelian gauge field strengths are defined by

$$W_\alpha=-\frac{1}{8}\overline D^2e^{-2V}D_\alpha e^{2V},\qquad \overline W_{\dot\alpha}=-\frac{1}{8}D^2e^{2V}\overline D_{\dot\alpha}e^{-2V}, \tag{4.47}$$

which, to first order, reduce to the abelian definition (4.37). Under gauge transformations they transform covariantly,

$$W_\alpha\to W'_\alpha=e^{-i\Lambda}W_\alpha e^{i\Lambda},\qquad \overline W_{\dot\alpha}\to\overline W'_{\dot\alpha}=e^{-i\Lambda^\dagger}W_\alpha e^{i\Lambda^\dagger}. \tag{4.48}$$

The component expansion of W_α^a in the Wess–Zumino gauge takes the form,

$$W_\alpha^a = -i\lambda_\alpha^a + \theta_\alpha D^a - \frac{i}{2}(\sigma^\mu \bar\sigma^\nu \theta)_\alpha F_{\mu\nu}^a + \theta^2 \sigma^\mu \nabla_\mu \bar\lambda^a \,, \qquad (4.49)$$

where

$$F_{\mu\nu}^a = \partial_\mu A_\nu^a - \partial_\nu A_\mu^a + f^{abc} A_\mu^b A_\nu^c \,, \qquad \nabla_\mu \bar\lambda^a = \partial_\mu \bar\lambda^a + f^{abc} A_\mu^b \bar\lambda^c \,, \quad (4.50)$$

and f^{abc} are the structure constants of the gauge group, $[T^a, T^b] = i f^{abc} T^c$. In the adjoint representation the generators T^a take the form: $(T^a)_{bc} = -i f^{abc}$.

The supersymmetric transformations in the non-abelian case are obtained following the same strategy as in the abelian case. Gauge restoring transformations have to be performed to remain in the Wess–Zumino gauge. The starting point is to consider the full transformation, the analog of (4.39) in the previous case. Owing to the exponential form of the gauge transformation (4.45) its form linearized in Λ is an infinite series in powers of V. However, in the Wess–Zumino gauge it becomes truncated at second order. The full transformation which one must take into consideration is,

$$\delta V^a = i(\xi Q - \bar\xi \bar Q)V^a + \frac{i}{2}(\Lambda - \Lambda^\dagger)^a - \frac{1}{2} f^{abc} V^b (\Lambda + \Lambda^\dagger)^c$$
$$- \frac{i}{6} f^{abc} f^{cde} V^b V^d (\Lambda - \Lambda^\dagger)^e \,. \qquad (4.51)$$

To maintain the conditions imposed on the Wess–Zumino gauge, $C^a = M^a = N^a = \chi^a = 0$, one needs to use all the component fields of Λ except the real part of its lower component. Denoting it by $v^a/2$ one finds the following set of transformations for the component fields that do not vanish in the Wess–Zumino gauge:

$$\delta A_\mu^a = \nabla_\mu v^a - i\sigma_{\alpha\dot\alpha}^\mu (\xi^\alpha \bar\lambda^{\dot\alpha a} + \bar\xi^{\dot\alpha} \lambda^{\alpha a}),$$
$$\delta\lambda_\alpha^a = f^{abc} v^b \lambda_\alpha^c + i\xi_\alpha D^a - i\xi_\beta (\sigma^{\mu\nu})_\alpha{}^\beta F_{\mu\nu}^a, \qquad (4.52)$$
$$\delta D^a = f^{abc} v^b D^c - \sigma_{\alpha\dot\alpha}^\mu (\xi^\alpha \nabla_\mu \bar\lambda^{\dot\alpha a} - \bar\xi^{\dot\alpha} \nabla_\mu \lambda^{\alpha a}),$$

being $\nabla_\mu v^a = \partial_\mu v^a + f^{abc} A_\mu^b v^c$.

The action is obtained in the same way as its abelian counterpart (4.44). As was pointed out there, it is convenient to introduce the complex parameter τ in (4.42). The action takes the form:

$$\frac{1}{4\pi}\mathrm{Im}\Big(\tau \int d^4x\, d^2\theta\, (W^{a\alpha} W_\alpha^a)\Big) =$$
$$\frac{1}{g^2}\int d^4x\, \Big(-\frac{1}{4} F^{a\mu\nu} F_{\mu\nu}^a - i\lambda^{a\alpha} \sigma_{\alpha\dot\alpha}^\mu \nabla_\mu \bar\lambda^{a\dot\alpha} + \frac{1}{2} D^a D^a\Big) \qquad (4.53)$$
$$- \frac{\Theta}{64\pi^2}\int d^4x\, \epsilon^{\mu\nu\rho\sigma} F_{\mu\nu}^a F_{\rho\sigma}^a \,.$$

We will now describe the couplings between the two $\mathcal{N} = 1$ multiplets studied in this section. To carry this out one must first assign a group representation to the chiral superfield, which will play the role of a matter field. Let us then assume that $\Phi = (\Phi_1, \ldots, \Phi_n)$ is a set of chiral superfields transforming in the following way under a gauge transformation,

$$\Phi \to e^{i\Lambda}\Phi, \qquad \Phi^\dagger \to \Phi^\dagger e^{-i\Lambda^\dagger}, \tag{4.54}$$

where $\Lambda = \Lambda^a T^a$ and T^a are chosen in the group representation R of dimension n carried by Φ. The superfield leading to the gauge invariant kinetic term takes the simple form,

$$\Phi^\dagger e^{2V} \Phi. \tag{4.55}$$

The most general action can now be written taking (4.28) as the starting point. First one must find out how to couple the vector multiplet keeping gauge invariance. The crucial observation is that if $K(\Phi, \Phi^\dagger)$ is a G-invariant function then it also remains invariant after replacing $\Phi^\dagger$ by $\Phi^\dagger e^{2V}$. Thus, taking into account (4.28) the most general action for a set of chiral superfields transforming in a representation R of the gauge group, coupled to a vector multiplet is:

$$\frac{1}{4\pi} \operatorname{Im} \left(\tau \int d^2\theta \, W^{a\alpha} W^a_\alpha \right) + \int d^2\theta d^2\overline{\theta} \, K(\Phi, \Phi^\dagger e^{2V})$$
$$+ \int d^2\theta \, \mathcal{W}(\Phi) + \int d^2\overline{\theta} \, \overline{\mathcal{W}}(\Phi^\dagger), \tag{4.56}$$

where $K(\Phi, \Phi^\dagger)$ and $\mathcal{W}(\Phi)$ are G-invariant functions. For the simple case in which the G-invariant function K takes the form $K(\Phi, \Phi^\dagger) = \Phi^\dagger \Phi$, after expanding this action in component fields one finds

$$\frac{1}{g^2} \int d^4x \left(-\frac{1}{4} F^{a\mu\nu} F^a_{\mu\nu} - i\lambda^{a\alpha} \sigma^\mu_{\alpha\dot\alpha} \nabla_\mu \overline{\lambda}^{a\dot\alpha} + \frac{1}{2} D^a D^a \right)$$
$$- \frac{\Theta}{64\pi^2} \int d^4x \, \epsilon^{\mu\nu\rho\sigma} F^a_{\mu\nu} F^a_{\rho\sigma}$$
$$+ \int d^4x \left((\nabla_\mu \phi)^{i\dagger} \nabla^\mu \phi^i - i\psi^i \sigma^\mu \nabla_\mu \overline{\psi}^i + F^{i\dagger} F^i + D^a \phi^{i\dagger} (T^a)_{ij} \phi^j \right.$$
$$+ i\sqrt{2} \, \phi^{i\dagger} (T^a)_{ij} \lambda^a \psi^j - i\sqrt{2} \, \overline{\psi}^i (T^a)_{ij} \phi^j \overline{\lambda}^a$$
$$\left. + \frac{\partial \mathcal{W}}{\partial \phi_i} F_i + \frac{\partial \overline{\mathcal{W}}}{\partial \phi_i^\dagger} F_i^\dagger - \frac{1}{2} \frac{\partial^2 \mathcal{W}}{\partial \phi_i \partial \phi_j} \psi_i \psi_j - \frac{1}{2} \frac{\partial^2 \overline{\mathcal{W}}}{\partial \phi_i^\dagger \partial \phi_j^\dagger} \overline{\psi}_i \overline{\psi}_j \right), \tag{4.57}$$

where $\mathcal{W}$ denotes the scalar component of the superpotential. In (4.57) the auxiliary fields F and D^a can be eliminated by using their field equations. The terms involving these fields thus give rise to the scalar potential

$$V = \sum_i \left| \frac{\partial \mathcal{W}}{\partial \phi_i} \right|^2 - \frac{1}{2} g^2 (\phi^\dagger T^a \phi)^2 . \tag{4.58}$$

The action (4.57) is invariant under supersymmetry and gauge transformations. These transformations are provided in (4.52) for the Yang–Mills fields. For the fields of the chiral superfields they turn out to be:

$$\begin{aligned}
\delta \phi^i &= \sqrt{2} \xi^\alpha \psi_\alpha^i - i v^a (T^a)_{ij} \phi^j, \\
\delta \psi_\alpha^i &= \sqrt{2} \xi_\alpha F^i + i\sqrt{2} \bar{\xi}^{\dot\alpha} \sigma_{\alpha\dot\alpha}^\mu \nabla \phi^i - i v^a (T^a)_{ij} \psi_\alpha^j, \\
\delta F^i &= i\sqrt{2} \bar{\xi}^{\dot\alpha} \sigma_{\alpha\dot\alpha}^\mu \nabla_\mu \psi^{\alpha i} + 2i \bar{\xi}_{\dot\alpha} \bar{\lambda}^{\dot\alpha a} (T^a)_{ij} \phi^j - i v^a (T^a)_{ij} F^j .
\end{aligned} \tag{4.59}$$

4.4. $\mathcal{N} = 2$ supersymmetric Yang–Mills theories

Theories with at least $\mathcal{N} = 2$ are of fundamental importance in the construction of topological quantum field theories. The twisting procedure that leads to these theories involves as its starting point a theory with at least two supersymmetries. In this section we will present the basic building block of this type of construction: the $\mathcal{N} = 2$ supersymmetric vector multiplet which leads to $\mathcal{N} = 2$ supersymmetric Yang–Mills theory. The formulation will be carried out in $\mathcal{N} = 1$ superspace, obtaining its simplest action. The construction of the most general model involving only this multiplet will then be carried out with the help of $\mathcal{N} = 2$ superspace. In the next section we will deal with the other $\mathcal{N} = 2$ supersymmetric multiplet, the hypermultiplet, and its coupling to the $\mathcal{N} = 2$ supersymmetric vector multiplet.

Let us consider a gauge group G and a series of real superfields, V^a, and chiral superfields, Φ^a, $a = 1, \ldots, \dim G$. To construct the action for this model we will start considering the action (4.56) for the simplest case in which $K(\Phi, \Phi^\dagger) = \Phi^{\dagger a} \Phi^a$ and $\mathcal{W} = 0$:

$$\frac{1}{4\pi} \mathrm{Im} \left(\tau \int d^2\theta \, W^{a\alpha} W_\alpha^a \right) + \int d^2\theta d^2\bar{\theta} \, \Phi^{\dagger a} (\mathrm{e}^{2V})_{ab} \Phi^b . \tag{4.60}$$

In this action the superfields W_α^a are the field strengths (4.47) corresponding to the non-abelian vector multiplet, and in $(\mathrm{e}^{2V})_{ab}$, V has the form $V = V^a T^a$ where T^a, $a = 1, \ldots, \dim G$, are in the adjoint representation, $(T^a)_{bc} = -i f_{abc}$.

To study further the theory we will consider it in the Wess–Zumino gauge. In this gauge the vector superfield V takes the group-valued form (4.36):

$$V = -\theta\sigma^\mu\bar{\theta}A_\mu - i\bar{\theta}^2\theta\lambda_2 + i\theta^2\bar{\theta}\bar{\lambda}^2 + \frac{1}{2}\theta^2\bar{\theta}^2 D, \qquad (4.61)$$

where we have made, for future convenience, the notational assignments $\lambda \to \lambda_2$ and $\bar{\lambda} \to \bar{\lambda}^2$. The chiral superfields Φ^a, $a = 1, \ldots, \dim G$, are assumed to have components ϕ^a, $-\lambda^a_{1\alpha}$ and F^a, and take the form (in the y, θ variables of (4.18)):

$$\Phi^a = \phi^a - \sqrt{2}\theta\lambda^a_1 + \theta^2 F^a,$$
$$\Phi^{\dagger a} = \phi^{\dagger a} - \sqrt{2}\bar{\theta\lambda}^{1a} + \bar{\theta}^2\overline{F}^a. \qquad (4.62)$$

We can now write the action (4.60) in terms of components. In doing this we will first rescale the superfield Φ as $\Phi \to \Phi/g$, and we will redefine the auxiliary field D as $D \to D + [\phi, \phi^\dagger]$. The component action then reads:

$$\frac{1}{g^2}\int d^4x\,\mathrm{Tr}\Big(\nabla_\mu\phi^\dagger\nabla^\mu\phi - i\lambda_1\sigma^\mu\nabla_\mu\bar{\lambda}^1 - i\lambda_2\sigma^\mu\nabla_\mu\bar{\lambda}^2 - \frac{1}{4}F_{\mu\nu}F^{\mu\nu}$$
$$+ \frac{1}{2}D^2 + |F|^2 - \frac{1}{2}[\phi, \phi^\dagger]^2 - i\sqrt{2}\lambda^\alpha_1[\phi^\dagger, \lambda_{2\alpha}] + i\sqrt{2}\bar{\lambda}^1_{\dot\alpha}[\bar{\lambda}^{2\dot\alpha}, \phi]\Big), \qquad (4.63)$$

plus the theta term in (4.57), which is topological. In obtaining (4.63) we have made the following choice for the normalization of the group generators in the adjoint representation: $\mathrm{Tr}\,(T^aT^b) = \delta^{ab}$. The action (4.63) is not manifestly $\mathcal{N} = 2$ supersymmetric. As described in section 5.1, the $\mathcal{N} = 2$ supersymmetric algebra has an internal $SU(2)_R$ symmetry. One would like to obtain a construction for the model under consideration in which this internal symmetry becomes manifest. Doing so we will easily find the full $\mathcal{N} = 2$ supersymmetric transformation of the component fields. The internal $SU(2)_R$ invariance is easily achieved after performing the following identification: the scalars ϕ^a and the gluons $A^a{}_\mu$ as $SU(2)_R$ singlets, the spinors λ_v, $v = 1, 2$, as elements of a $SU(2)_R$ doublet and, finally, the auxiliary fields as elements of a real $SU(2)_R$ triplet:

$$D^{vw} = \begin{pmatrix} \sqrt{2}F & iD \\ iD & \sqrt{2}\overline{F} \end{pmatrix}. \qquad (4.64)$$

In doublets and triplets the $SU(2)_R$ indices are raised and lowered with the antisymmetric matrices ϵ_{vw} and ϵ^{vw} ($\epsilon_{12} = 1$, $\epsilon_{vu}\epsilon^{uw} = \delta^w_v$). Notice that $D^{vw} = D^*_{vw}$. The full $\mathcal{N} = 2$ supersymmetric transformations are obtained by making covariant the $\mathcal{N} = 1$ transformations (4.52) and (4.59) after labelling as ξ_1 the supersymmetry parameter ξ and introducing an additional parameter ξ_2. These are elements of a $SU(2)_R$ doublet. The final form of the $\mathcal{N} = 2$ supersymmetric transformations is:

$$\delta\phi = \sqrt{2}\epsilon^{vw}\xi_v\lambda_w,$$

$$\delta A_\mu = i\xi_v\sigma_\mu\overline{\lambda}^v - i\lambda_v\sigma_\mu\overline{\xi}^v,$$

$$\delta\lambda_{v\alpha} = D_v{}^w\xi_{w\alpha} - i\xi_{v\alpha}[\phi,\phi^\dagger] - i\sigma^{\mu\nu}{}_\alpha{}^\beta\xi_{v\beta}F_{\mu\nu}$$
$$+ i\sqrt{2}\epsilon_{vw}\sigma^\mu_{\alpha\dot\alpha}\overline{\xi}^{w\dot\alpha}\nabla_\mu\phi,$$

$$\delta D^{vw} = 2i\overline{\xi}^{(v}\overline{\sigma}^\mu\nabla_\mu\lambda^{w)} + 2i\nabla_\mu\overline{\lambda}^{(v}\overline{\sigma}^\mu\xi^{w)}$$
$$+ 2i\sqrt{2}\xi^{(v}[\lambda^{w)},\phi^\dagger] + 2i\sqrt{2}\overline{\xi}^{(v}[\overline{\lambda}^{w)},\phi].$$

$$(4.65)$$

In these equations $\xi_v{}^\alpha$ and $\overline{\xi}^v{}_{\dot\alpha}$ are fermionic parameters, and the δ-transformation is generated by the charge

$$i\epsilon^{vw}\xi_v{}^\alpha Q_{\alpha w} + i\epsilon_{vw}\overline{\xi}^v{}_{\dot\alpha}\overline{Q}^{\dot\alpha w}.\qquad(4.66)$$

The action (4.63) now reads

$$\frac{1}{g^2}\int d^4x\,\mathrm{Tr}\Big(\nabla_\mu\phi^\dagger\nabla^\mu\phi - i\lambda_v\sigma^\mu\nabla_\mu\overline{\lambda}^v - \frac{1}{4}F_{\mu\nu}F^{\mu\nu} + \frac{1}{4}D_{vw}D^{vw}$$
$$- \frac{1}{2}[\phi,\phi^\dagger]^2 - \frac{i}{\sqrt{2}}\epsilon^{vw}\lambda_v{}^\alpha[\phi^\dagger,\lambda_{w\alpha}] - \frac{i}{\sqrt{2}}\epsilon_{vw}\overline{\lambda}^v{}_{\dot\alpha}[\overline{\lambda}^{w\dot\alpha},\phi]\Big).$$

$$(4.67)$$

In (4.65) and (4.67) the internal symmetry $SU(2)_R$ is manifest. The above action also has a classical $U(1)_R$ symmetry. Introducing φ as a parameter for this symmetry the component fields have the following transformations and $U(1)_R$ charges q_R:

$$
\begin{array}{llll}
A_\mu \to A_\mu, & q_R = 0; & D_{vw} \to D_{vw}, & q_R = 0; \\
\lambda_{v\alpha} \to e^{i\varphi}\lambda_{v\alpha}, & q_R = 1; & \phi \to e^{2i\varphi}\phi, & q_R = 2; \\
\overline{\lambda}^v{}_{\dot\alpha} \to e^{-i\varphi}\overline{\lambda}^v{}_{\dot\alpha}, & q_R = -1; & \phi^\dagger \to e^{-2i\varphi}\phi^\dagger, & q_R = -2.
\end{array}
$$

$$(4.68)$$

These transformations can also be defined at the level of the $\mathcal{N} = 1$ superfields:

$$W_\alpha \to e^{-i\varphi}W_\alpha(e^{i\varphi}\theta), \qquad \Phi \to e^{-2i\varphi}\Phi(e^{i\varphi}\theta). \qquad (4.69)$$

In general, extended superfields are not very useful for defining supersymmetric theories with extended supersymmetries. Extended superfields involve too many component fields, and it is difficult to define simple constraints for dealing with irreducible representations. An exception to this general property is starred by the $\mathcal{N} = 2$ vector multiplet. In fact, making use of the $\mathcal{N} = 2$ superspace formalism one derives very easily the most general form of the action for this multiplet. We will not introduce here the basic notions of extended superspace in order to derive this result. We will simply state it and refer the reader to standard references where this is explained in detail. It turns out that the most general $\mathcal{N} = 2$ supersymmetric action involving only the $\mathcal{N} = 2$ vector multiplet can be written in terms of a single gauge-invariant holomorphic function $\mathcal{F}(\Phi)$ called the *prepotential*. For the case of an abelian gauge group, which is the situation of interest in this book, this action takes the following form in $\mathcal{N} = 1$ superspace:

$$\frac{1}{4\pi} \operatorname{Im} \left(\int d^4\theta \, \frac{\partial \mathcal{F}(\Phi)}{\partial \Phi} \overline{\Phi} + \int d^2\theta \, \frac{1}{2} \frac{\partial^2 \mathcal{F}}{\partial \Phi^2} W^\alpha W_\alpha \right). \tag{4.70}$$

Notice that this reduces to (4.60) for the abelian case after considering $\mathcal{F}(\Phi) = \frac{\tau}{2}\Phi^2$.

4.5. $\mathcal{N} = 2$ supersymmetric hypermultiplets

The $\mathcal{N} = 2$ supersymmetry generalization of the chiral multiplet is the hypermultiplet. In terms of $N = 1$ superfields it consists of two chiral multiplets. Adequate choices of the component fields of these chiral multiplets leads us to discover the internal $SU(2)_R$ symmetry. It turns out that these build a $SU(2)_R$ doublet of complex scalar fields, two singlet spinors, and a $SU(2)_R$ doublet of auxiliary fields. The components of the two chiral multiplets, Q and $\widetilde{Q}$, are defined as (in the y, θ variables of (4.18)):

$$\begin{aligned} Q &= q_2 + \sqrt{2}\theta\psi + \theta^2 F_1, \\ \widetilde{Q} &= i(q^{\dagger 1} + \sqrt{2}\theta\chi + \theta^2 F^{\dagger 2}), \end{aligned} \tag{4.71}$$

and the action is

$$\int d^4x d^2\theta d^2\overline{\theta}(Q^\dagger Q + \widetilde{Q}^\dagger \widetilde{Q}) - im \int d^4x d^2\theta \, Q\widetilde{Q} + im \int d^4x d^2\overline{\theta} \, Q^\dagger \widetilde{Q}^\dagger. \tag{4.72}$$

The free action in terms of the component fields of the hypermultiplet takes the form:

$$\begin{aligned} \int d^4x \, \Big(&\partial_\mu q^{\dagger 1}\partial^\mu q_1 + \partial_\mu q^{\dagger 2}\partial^\mu q_2 - i\psi\sigma^\mu\partial_\mu\overline{\psi} - i\chi\sigma^\mu\partial_\mu\overline{\chi} \\ &+ F^{\dagger 1}F_1 + F^{\dagger 2}F_2 + mF^{\dagger 1}q_1 + mF^{\dagger 2}q_2 + mF_1q^{\dagger 1} + mF_2q^{\dagger 2} \\ &- m\overline{\psi}\overline{\chi} - m\psi\chi \Big). \end{aligned} \tag{4.73}$$

This action has a manifest $SU(2)_R$ invariance, with two doublets q_w and F_w, and can be written as

$$\int d^4x \left(\partial_\mu q^{\dagger w} \partial^\mu q_w - i\psi\sigma^\mu\partial_\mu\overline{\psi} - i\chi\sigma^\mu\partial_\mu\overline{\chi} - m\overline{\psi}\overline{\chi} - m\psi\chi \right.$$
$$\left. + F^{\dagger w}F_w + mF^{\dagger w}q_w + mF_w q^{\dagger w} \right). \tag{4.74}$$

The usual $\mathcal{N} = 1$ supersymmetric transformations (4.21) can be extended to $\mathcal{N} = 2$ supersymmetry carrying out the corresponding covariantizations and introducing an additional parameter ξ_2 as in (4.65):

$$\delta q_v = \sqrt{2}\xi_v\psi + \sqrt{2}\epsilon_{vw}\overline{\xi}^w\overline{\chi},$$
$$\delta\psi = \sqrt{2}\epsilon^{vw}\xi_v F_w + i\sqrt{2}\sigma^\mu\overline{\xi}^w\partial_\mu q_w,$$
$$\delta\chi = \sqrt{2}\xi_w F^{\dagger w} - i\sqrt{2}\epsilon_{vw}\sigma^\mu\overline{\xi}^v\partial_\mu q^{\dagger w}, \tag{4.75}$$
$$\delta F_v = -i\sqrt{2}\partial_\mu\psi\sigma^\mu\epsilon_{vw}\overline{\xi}^w + i\sqrt{2}\xi_v\sigma^\mu\partial_\mu\overline{\chi}.$$

An analysis of the algebra of these transformations shows that it possesses a central charge as the one contained in the general form of the supersymmetry algebra (4.2). The action of the central charge on the component fields turns out to be:

$$\delta_Z q_v \sim F_v,$$
$$\delta_Z \psi_\alpha \sim \sigma^\mu_{\alpha\dot\alpha}\partial_\mu\overline{\psi}^{\dot\alpha},$$
$$\delta_Z \chi_\alpha \sim \sigma^\mu_{\alpha\dot\alpha}\partial_\mu\overline{\chi}^{\dot\alpha}, \tag{4.76}$$
$$\delta_Z F_v \sim \partial^\mu\partial_\mu\overline{q}_v.$$

These transformations show that the value of the central charge is proportional to the mass hypermultiplet m when taking into account the field equations of the component fields emerging from the action (4.74). Thus there are no central charges in the massless case. Notice, however, that even with central charges present the model possesses the internal symmetry $SU(2)_R$. This because in $\mathcal{N} = 2$ the reduction $U(N) \to USp(N)$ in the presence of central charges is just $U(2) \to SU(2)$ since any $SU(2)$ transformation leaves invariant any two-form $Z_{uv} \propto \epsilon_{uv}$. On the other hand, the theory does not have a $U(1)_R$ symmetry when central charges are present. The part of the algebra (4.2), $\{Q_{u\alpha}, Q_{v\beta}\} = C_{\alpha\beta}Z_{uv}$, implies that the $U(1)_R$ charge of the fermionic generators $Q_{u\alpha}$ vanishes. One finds easily that a charge assignment at the level of superfield as in (4.69) is not possible.

In the massless case the theory possesses a $U(1)_R$ symmetry with the following charge assignment for the fields:

$$\psi \to e^{i\varphi}\psi, \qquad q_v \to q_v,$$
$$\chi \to e^{i\varphi}\chi, \qquad F_v \to e^{iq_F\varphi}F_v. \tag{4.77}$$

The $U(1)_R$ charge for the field q has been selected to match the choice made for the $\mathcal{N}=2$ vector multiplet. The charge q_F of F_v is arbitrary.

4.6. $\mathcal{N}=2$ supersymmetric Yang–Mills theories with matter

In this section we study the coupling of the $\mathcal{N}=2$ supersymmetric matter hypermultiplet to the $\mathcal{N}=2$ supersymmetric vector multiplet. From the point of view of $\mathcal{N}=1$ superspace the theory consists of two chiral superfields Q and $\widetilde{Q}$ in conjugate representations R and $\widetilde{R}$ of the gauge group and a real superfield V together with a chiral superfield Φ in the adjoint representation. The action is a particular case of (4.56), after taking the following Kähler potential and superpotential:

$$K = \Phi^\dagger\Phi + Q^\dagger Q + \widetilde{Q}\widetilde{Q}^\dagger, \qquad W = \sqrt{2}\widetilde{Q}\Phi Q - im\widetilde{Q}Q. \tag{4.78}$$

The full action then reads

$$\frac{1}{4\pi}\,\mathrm{Im}\left(\tau\int d^2\theta\, W^{a\alpha}W_\alpha^a\right) + \int d^4x d^2\theta d^2\overline{\theta}\,\Phi^\dagger e^{2V}\,\Phi$$
$$+ \int d^4x d^2\theta d^2\overline{\theta}\,(Q^\dagger e^{2V}Q + \widetilde{Q}e^{-2V}\widetilde{Q}^\dagger) + \sqrt{2}\int d^4x d^2\theta\,\widetilde{Q}\Phi Q \tag{4.79}$$
$$+ \sqrt{2}\int d^4x d^2\overline{\theta}\, Q^\dagger \Phi^\dagger \widetilde{Q}^\dagger - im\int d^4x d^2\theta\,\widetilde{Q}Q + im\int d^4x d^2\overline{\theta}\, Q^\dagger\widetilde{Q}^\dagger.$$

This action can be written in terms of component fields using the expansions (4.61), (4.62) and (4.71). In order to find a manifest $SU(2)_R$ invariant expression, it is convenient to redefine the auxiliary fields involved in those expansions in the following way:

$$\widehat{F}_1 = F_1 + mq_1 + i\sqrt{2}\phi^\dagger q_1,$$
$$\widehat{F}_2 = F_2 + mq_2 - i\sqrt{2}\phi q_2,$$
$$\widehat{D}^a = D^a + [\phi, \phi^\dagger]^a - q^{\dagger 1}T^a q_1 + q^{\dagger 2}T^a q_2, \tag{4.80}$$
$$\widehat{F}^a = F^a - i\sqrt{2}q^{\dagger 2}T^a q_1,$$
$$\widehat{\overline{F}}^a = \overline{F}^a + i\sqrt{2}q^{\dagger 1}T^a q_2.$$

The resulting action in terms of component fields turns out to be:

$$
\int d^4x \, \mathrm{Tr}\Big(\nabla_\mu \phi^\dagger \nabla^\mu \phi - i\lambda_w \sigma^\mu \nabla_\mu \overline{\lambda}^w - \frac{1}{4} F_{\mu\nu} F^{\mu\nu} + \frac{1}{4} D_{vw} D^{vw}
$$
$$
- \frac{1}{2}[\phi, \phi^\dagger]^2 - \frac{i}{\sqrt{2}} \epsilon^{vw} \lambda_v^\alpha [\phi^\dagger, \lambda_{w\alpha}] - \frac{i}{\sqrt{2}} \epsilon_{vw} \overline{\lambda}^v{}_{\dot\alpha} [\overline{\lambda}^{w\dot\alpha}, \phi]\Big)
$$
$$
+ \int d^4x \Big(\nabla_\mu q^{\dagger w} \nabla^\mu q_w - i\psi \sigma^\mu \partial_\mu \overline{\psi} - i\chi \sigma^\mu \partial_\mu \overline{\chi} - m\overline{\psi}\,\overline{\chi} - m\psi\chi + m^2 q^{\dagger w} q_w
$$
$$
+ F^{\dagger v} F_v + i\sqrt{2} q^{\dagger v} \lambda_v \psi + i\sqrt{2} \epsilon_{vw} q^{\dagger v} \overline{\lambda}^w \overline{\chi} - i\sqrt{2} \epsilon^{vw} \chi \lambda_v q_w
$$
$$
- i\sqrt{2}\overline{\psi} \overline{\lambda}^v q_v + i\sqrt{2}\,\chi^\alpha \phi \psi_\alpha - i\sqrt{2}\,\overline{\psi}_{\dot\alpha} \phi^\dagger \overline{\chi}^{\dot\alpha} - q^{\dagger w}\{\phi, \phi^\dagger\} q_w
$$
$$
+ im\sqrt{2} q^{\dagger w} \phi^\dagger q_w - im\sqrt{2} q^{\dagger w} \phi q_w + q^{\dagger(v} T^a q^{w)} q^\dagger{}_{(v} T^a q_{w)}\Big),
$$

$$(4.81)$$

where we have suppressed the hats over the auxiliary fields.

The $\mathcal{N} = 2$ supersymmetric transformations take the following form. For the gauge fields,

$$
\delta\phi = \sqrt{2}\epsilon^{vw} \xi_v \lambda_w,
$$
$$
\delta A_\mu = i\xi_v \sigma_\mu \overline{\lambda}^v - i\lambda_v \sigma_\mu \overline{\xi}^v,
$$
$$
\delta\lambda^a{}_{v\alpha} = D^a{}_v{}^w \xi_{w\alpha} - i\xi_{v\alpha}[\phi, \phi^\dagger]^a - i\sigma^{\mu\nu}{}_\alpha{}^\beta \xi_{v\beta} F^a_{\mu\nu}
$$
$$
+ i\sqrt{2}\epsilon_{vw} \sigma^\mu_{\alpha\dot\alpha} \overline{\xi}^{w\dot\alpha} \nabla_\mu \phi^a + 2iq^\dagger{}_{(v} T^a q_{w)} \xi^w{}_\alpha,
$$

$$(4.82)$$

$$
\delta D^a{}^{vw} = 2i\overline{\xi}^{(v} \overline{\sigma}^\mu \nabla_\mu \overline{\lambda}^{a\ w)} + 2i\nabla_\mu \overline{\lambda}^a{}^{(v} \overline{\sigma}^\mu \xi^{w)} + 2\sqrt{2}i\xi^{(v}[\lambda^{w)}, \phi^\dagger]^a
$$
$$
+ 2\sqrt{2}i\overline{\xi}^{(v}[\overline{\lambda}^{w)}, \phi]^a + 2q^{\dagger(v} T^a \psi \xi^{w)} + 2\sqrt{2}i\xi^{(v}\overline{\psi} T^a q^{w)}
$$
$$
+ 2\sqrt{2}iq^{\dagger(v} T^a \overline{\chi}\,\overline{\xi}^{w)} - 2\sqrt{2}i\xi^{(v}\chi T^a q^{w)},
$$

and for the matter fields,

$$
\delta q_v = \sqrt{2}\xi_v \psi + \sqrt{2}\epsilon_{vw} \overline{\xi}^w \overline{\chi},
$$
$$
\delta\psi = \sqrt{2}\epsilon^{vw} \xi_v F_w + i\sqrt{2}\sigma^\mu \overline{\xi}^w \nabla_\mu q_w - \sqrt{2}m\epsilon^{vw} \xi_v q_w - 2i\epsilon^{vw} \xi_v \phi^\dagger q_w,
$$
$$
\delta\chi = \sqrt{2}\xi_w F^{\dagger w} - i\sqrt{2}\epsilon_{vw} \sigma^\mu \overline{\xi}^v \nabla_\mu q^{\dagger w} - \sqrt{2}mq^{\dagger v} \xi_v - 2i\xi_v q^{\dagger v} \phi^\dagger,
$$

$$(4.83)$$

$$
\delta F_v = -i\sqrt{2}\nabla_\mu \psi \sigma^\mu \epsilon_{vw} \overline{\xi}^w + i\sqrt{2}\xi_v \sigma^\mu \nabla_\mu \overline{\chi} + \sqrt{2}m\xi_v \psi - 2i\xi_v \phi \psi
$$
$$
+ 2i\xi_v \epsilon^{wz} \lambda_w q_z + \sqrt{2}m\epsilon_{vw} \overline{\xi}^w \overline{\chi} + 2i\epsilon_{vw} \overline{\xi}^w \phi^\dagger \overline{\chi} + 2i\epsilon_{vw} \overline{\xi}^w \overline{\lambda}^z q_z.
$$

Finally, in the massless theory, $m = 0$, there is a $U(1)_R$ symmetry once the auxiliary fields F_v have been integrated out:

$$
\begin{aligned}
A_\mu &\to A_\mu, & q_R &= 0, \\
\lambda_{v\alpha} &\to e^{i\varphi}\lambda_{v\alpha}, & q_R &= 1, \\
\overline{\lambda}^v{}_{\dot\alpha} &\to e^{-i\varphi}\overline{\lambda}^v{}_{\dot\alpha}, & q_R &= 1, \\
\psi &\to e^{-i\varphi}\psi, & q_R &= -1, \\
\chi &\to e^{-i\varphi}\chi, & q_R &= -1.
\end{aligned}
\qquad
\begin{aligned}
D_{vw} &\to D_{vw}, & q_R &= 0, \\
\phi &\to e^{2i\varphi}\phi, & q_R &= 2, \\
\phi^\dagger &\to e^{-2i\varphi}\phi^\dagger, & q_R &= -2, \\
q_v &\to q_v, & q_R &= 0,
\end{aligned}
$$

$$(4.84)$$

It is not possible to build an off-shell formulation of the massless theory in which the $U(1)_R$ symmetry is present. This fact will have important consequences in the study of the twisted versions of this theory. In twisted theories the symmetry which is equivalent to the $U(1)_R$ symmetry in the untwisted theories plays an important role off-shell and should be maintained. As will be described in Chapter 5, this will be achieved by modifying the auxiliary field content in the twisted theory.

Bibliographical notes

- Two standard references on supersymmetry and $\mathcal{N}=1$ supersymmetric field theories are [40] and [41].

- A useful summary for the purpose of this book can be found in [42].

Chapter 5

Topological Quantum Field Theories in Four Dimensions

In this chapter we introduce topological quantum field theories (TQFTs) and we give a brief general overview of their properties, focusing on the so called theories of the Witten or cohomological type. We then explain the twisting procedure, which produces topological quantum field theories from $\mathcal{N} = 2$ supersymmetric theories, and put it into practice with the examples of the previous chapter. In this way we shall be able to construct a topological quantum field theory in four-dimensions which gives a physical realization of Donaldson theory, the so called Donaldson–Witten theory.

5.1. Basic properties of topological quantum field theories

Let X be a manifold with the Riemannian metric $g_{\mu\nu}$, and let us consider a quantum field theory defined on X. In general the partition function and correlation functions of this theory will depend on the background metric. We will say that a quantum field theory is *topological* if there exists a set of operators in the theory (that we shall call *topological observables*) such that their correlation functions do not depend on the metric. If we denote these operators by $\mathcal{O}_i$ (where i is a label) then

$$\frac{\delta}{\delta g_{\mu\nu}} \langle \mathcal{O}_{i_1} \cdots \mathcal{O}_{i_n} \rangle = 0. \tag{5.1}$$

There are two different types of TQFTs. In the TQFTs of the *Schwarz type* one defines the relevant ingredients in the theory (the action and the observables) without using the metric of the manifold. This guarantees topological invariance as a classical symmetry of the theory, and in some cases the quantization procedure can be seen to preserve this classical symmetry so that one has (5.1). The most important example of a TQFT of Schwarz type is Chern–Simons gauge theory, introduced by Witten in 1988.

In the TQFTs of the *Witten type* (also called *cohomological* TQFTs) the action and the operators may depend on the metric, but the theory has an underlying scalar symmetry δ acting on the fields ϕ_i in such a way that the correlation functions of the theory do not depend on the background metric. This is achieved as follows. Since δ is a symmetry the action of the theory satisfies $\delta S(\phi_i) = 0$. If the energy-momentum tensor of the theory $T_{\mu\nu} = (\delta/\delta g^{\mu\nu})S(\phi_i)$ can be written as

$$T_{\mu\nu} = -i\delta G_{\mu\nu}, \tag{5.2}$$

where $G_{\mu\nu}$ is some tensor, then (5.1) holds for any operator $\mathcal{O}$ which is δ-invariant. This is because

$$\frac{\delta}{\delta g^{\mu\nu}}\langle \mathcal{O}_{i_1}\mathcal{O}_{i_2}\cdots\mathcal{O}_{i_n}\rangle = \langle \mathcal{O}_{i_1}\mathcal{O}_{i_2}\cdots\mathcal{O}_{i_n}T_{\mu\nu}\rangle$$

$$= -i\langle \mathcal{O}_{i_1}\mathcal{O}_{i_2}\cdots\mathcal{O}_{i_n}\delta G_{\mu\nu}\rangle = \pm i\langle\delta(\mathcal{O}_{i_1}\mathcal{O}_{i_2}\cdots\mathcal{O}_{i_n}G_{\mu\nu})\rangle = 0. \tag{5.3}$$

The symmetry δ is usually called a *topological symmetry*. In (5.3) we have assumed that the symmetry δ is not anomalous, so that it is a full symmetry of the quantum theory. Moreover, we have 'integrated by parts' in field space, therefore we have assumed that there are no contributions coming from boundary terms. In some situations these assumptions do not hold, and the theory is then not strictly topological. However, in most of the interesting cases the resulting dependence on the metric is mild and under control. We will see a very explicit example of this in Donaldson theory on manifolds with $b_2^+ = 1$.

In a cohomological theory the observables are the δ-invariant operators. On the other hand, operators which are δ-exact decouple from the theory, since their correlation functions vanish. We will then restrict the set of observables to the cohomology of δ:

$$\mathcal{O} \in \frac{\text{Ker}\,\delta}{\text{Im}\,\delta}. \tag{5.4}$$

In all known examples of cohomological TQFTs, δ is a Grasmannian symmetry. Since, on the other hand, δ is a scalar symmetry we see that topological quantum field theories of Witten type violate the spin-statistics theorem. In general δ is is not nilpotent, and one has

$$\delta^2 = Z, \tag{5.5}$$

where Z is a certain transformation in the theory. It can be a local transformation (a gauge transformation) or a global transformation (for example, a global $U(1)$ symmetry). The appropriate framework for analysing the observables is then *equivariant cohomology*, and for consistency one has to consider only operators that are invariant under the transformation generated by Z (for example, gauge invariant operators). Equivariant cohomology turns out to be a very natural language to describe TQFTs with local and global symmetries, and we shall present it in some detail when we consider the geometric formulation of cohomological field theories in Chapter 6.

The structure of topological quantum field theories of the Witten type leads immediately to a general version of the Donaldson map. Remember that, starting with the curvature of the universal bundle, this map associates cohomology classes in the instanton moduli space to homology classes in the four-manifold. Let us now assume that we have found an operator $\phi^{(0)}$ which is in the cohomology of δ, as well as a series of operators $\phi^{(n)}$, $n = 1, \ldots, \dim X$, which are differential forms of degree n on X such that

$$d\phi^{(n)} = \delta\phi^{(n+1)}, \quad n \geq 0. \tag{5.6}$$

In this equation d denotes the exterior derivative on X. The operators $\phi^{(n)}$ are called the *topological descendants* of $\phi^{(0)}$. It is easy to see that the operator

$$W_{\phi^{(0)}}^{(\gamma_n)} = \int_{\gamma_n} \phi^{(n)}, \tag{5.7}$$

where $\gamma_n \in H_n(X)$, is a topological observable:

$$\delta W_{\phi^{(0)}}^{(\gamma_n)} = \int_{\gamma_n} \delta\phi^{(n)} = \int_{\gamma_n} d\phi^{(n-1)} = \int_{\partial\gamma_n} \phi^{(n-1)} = 0, \tag{5.8}$$

since $\partial\gamma_n = 0$. Similarly, it is easy to show that if γ_n is trivial in homology (*i.e.*, if it is ∂-exact) then $W_{\phi^{(0)}}^{(\gamma_n)}$ is δ-exact. The equations (5.6) are called *topological descent equations*. The conclusion of this analysis is that, given a (scalar) topological observable $\phi^{(0)}$ and a solution to the descent equations (5.6), one can construct a family of topological observables

$$W_{\phi^{(0)}}^{(\gamma_{i_n})}, \quad i_n = 1, \ldots, b_n; \quad n = 1, \ldots, \dim X, \tag{5.9}$$

in one-to-one correspondence with homology classes of space-time. This descent procedure is the analog of the Donaldson map in Donaldson–Witten and Seiberg–Witten theory.

It is easy to see that in any theory in which (5.2) is satisfied there is a simple procedure for constructing a solution to (5.6) given a scalar observable $\phi^{(0)}$. If (5.2) holds then one has

$$P_\mu = T_{0\mu} = -i\delta G_\mu, \tag{5.10}$$

where

$$G_\mu \equiv G_{0\mu}. \tag{5.11}$$

Since δ is a Grassmannian symmetry, G_μ is an anticommuting operator and a one-form in space-time. If we are given a δ-invariant operator $\phi^{(0)}(x)$ we can use (5.11) to construct the topological descendants

$$\phi^{(n)}_{\mu_1\mu_2\ldots\mu_n}(x) = G_{\mu_1} G_{\mu_2} \cdots G_{\mu_n} \phi^{(0)}(x), \quad n = 1, \ldots, \dim X. \tag{5.12}$$

On the other hand, since the G_{μ_i} anticommute

$$\phi^{(n)} = \frac{1}{n!} \phi^{(n)}_{\mu_1\mu_2\ldots\mu_n} dx^{\mu_1} \wedge \cdots \wedge dx^{\mu_n} \tag{5.13}$$

is an n-form on X. By using (5.10) the δ-invariance of $\phi^{(0)}$, as well as $P_\mu = -i\partial_\mu$, one can easily check that these forms satisfy the descent equations (5.6). This solution to (5.6) is usually called the *canonical solution* of the descent equations.

5.2. Twist of $\mathcal{N} = 2$ supersymmetry

In the early eighties Witten noticed that supersymmetry has a deep relation to topology. The simplest example of such a relation is supersymmetric quantum mechanics, which provides a physical reformulation (and, in fact, a refinement) of Morse theory. Another examples are $\mathcal{N} = 2$ supersymmetric theories in two and four dimensions. In 1988 Witten discovered that, by changing the coupling to gravity of the fields in an $\mathcal{N} = 2$ supersymmetric theory in two or four dimensions, a TQFT theory of cohomological type was obtained. This redefinition of the theory is called *twisting*. We are now going to explain in some detail how this works in the four-dimensional case.

Our starting point is the Euclidean version of the $\mathcal{N} = 2$ supersymmetry algebra (with no central charges) which is easily obtained from (4.2), (4.5) and (4.7):

$$\{Q_{\alpha v}, \overline{Q}_{\dot\beta w}\} = 2\epsilon_{vw}\sigma^\mu_{\alpha\dot\beta}P_\mu, \qquad\qquad \{Q_{\alpha v}, Q_{\beta w}\} = 0,$$

$$[P_\mu, Q_{\alpha v}] = 0, \qquad\qquad\qquad\qquad [P_\mu, \overline{Q}_{\dot\alpha v}] = 0,$$

$$[M_{\mu\nu}, Q_{\alpha v}] = -(\sigma_{\mu\nu})_\alpha{}^\beta Q_{\beta v}, \qquad [M_{\mu\nu}, \overline{Q}^{\dot\alpha v}] = -(\overline{\sigma}_{\mu\nu})^{\dot\alpha}{}_{\dot\beta}\overline{Q}^{\dot\beta v}, \quad (5.14)$$

$$[Q_{\alpha v}, B^a] = -\frac{1}{2}(\tau^a)_v{}^w Q_{\alpha w}, \qquad [\overline{Q}_{\dot\alpha}{}^v, B^a] = \frac{1}{2}\overline{Q}_{\dot\alpha}{}^w(\tau^a)_w{}^v,$$

$$[Q_{\alpha v}, R] = Q_{\alpha v}, \qquad\qquad\qquad [\overline{Q}_{\dot\alpha v}, R] = -\overline{Q}_{\dot\alpha v}.$$

Here $v, w \in \{1, 2\}$ are $SU(2)_R$ indices and $-\frac{1}{2}(\tau^a)_v{}^w$, $a = 1, 2, 3$, are the matrices $(b^a)_v{}^w$ in (4.5), $i.e.$, the generators of this internal group, $(b^a)_v{}^w = -\frac{1}{2}(\tau^a)_v{}^w$. The matrices τ^a are the Pauli matrices,

$$\tau^1 = \begin{pmatrix} 0 & 1 \\ 1 & 0 \end{pmatrix}, \qquad \tau^2 = \begin{pmatrix} 0 & -i \\ i & 0 \end{pmatrix}, \qquad \tau^3 = \begin{pmatrix} 1 & 0 \\ 0 & -1 \end{pmatrix}. \quad (5.15)$$

In these relations we have performed the replacement $p^0 = -ip^4$ and all the matrices σ^μ, $\sigma^{\mu\nu}$ and $\overline{\sigma}^{\mu\nu}$ correspond to the Euclidean counterparts, as defined in Appendix A. The total symmetry group of the theory is

$$\mathcal{H} = SU(2)_+ \times SU(2)_- \times SU(2)_R \times U(1)_R, \quad (5.16)$$

being $\mathcal{K} = SU(2)_+ \times SU(2)_-$ the rotation group, and $SU(2)_R \times U(1)_R$ the internal group. Under this group the supersymmetry generators $Q_{\alpha v}$ and $\overline{Q}_{\dot\alpha v}$ transform as $(\mathbf{0}, \mathbf{2}, \mathbf{2})^1$ and $(\mathbf{2}, \mathbf{0}, \mathbf{2})^{-1}$, respectively. The generator of the rotation group, $M_{\mu\nu}$, can be decomposed in terms of bispinors as described in Appendix A:

$$M_{\mu\nu} \to M_{\alpha\dot\alpha, \beta\dot\beta} = \epsilon_{\alpha\beta}\overline{M}_{\dot\alpha\dot\beta} + \epsilon_{\dot\alpha\dot\beta}M_{\alpha\beta}, \quad (5.17)$$

so that $\overline{M}_{\dot\alpha\dot\beta}$ are the generators of $SU(2)_+$ and $M_{\alpha\beta}$ the generators of $SU(2)_-$. Their action on the supersymmetry generators $Q_{\alpha v}$ and $\overline{Q}_{\dot\alpha v}$ turns out to be:

$$[M_{\alpha\beta}, Q_{\delta v}] = \epsilon_{\delta(\alpha}Q_{\beta)v}, \qquad [M_{\alpha\beta}, \overline{Q}_{\dot\alpha v}] = 0,$$
$$[\overline{M}_{\dot\alpha\dot\beta}, Q_{\delta v}] = 0, \qquad\qquad [\overline{M}_{\dot\alpha\dot\beta}, \overline{Q}_{\dot\delta v}] = \epsilon_{\dot\delta(\dot\alpha}\overline{Q}_{\dot\beta)v}. \quad (5.18)$$

The action of the generators B^a of the internal group $SU(2)_R$ can be rewritten using their form in terms of the symmetric bispinor B^{vw},

$$[B^{vw}, Q_\alpha^u] = \epsilon^{u(v}Q_\alpha^{w)}, \qquad [B^{vw}, \overline{Q}_{\dot\alpha}^u] = -\epsilon^{u(v}\overline{Q}_{\dot\alpha}^{w)}. \quad (5.19)$$

The twisting procedure consists of redefining the coupling of the theory to gravity, which is carried out by redefining the spins of the fields. To do this we couple the fields to the $SU(2)_+$ spin connection according to their isospin, *i.e.*, according to the way they transform under the internal group $SU(2)_R$. This means that we identify the isospin $SU(2)_R$ indices v, w with the $SU(2)_+$ indices $\dot\alpha$, $\dot\beta$. Therefore we define a generator $M'_{\dot\alpha\dot\beta}$ as follows:

$$M'_{\dot\alpha\dot\beta} = M_{\dot\alpha\dot\beta} - B_{\dot\alpha\dot\beta}. \tag{5.20}$$

We then identify a new rotation group $\mathcal{K}' = SU'(2)_+ \otimes SU(2)_-$, where $SU'(2)_+$ is generated by M'. Under the twisting the isospin index v becomes a spinorial index, $\overline{Q}_{\dot\alpha v} \to \overline{Q}_{\dot\alpha\dot\beta}$ and $Q_{\alpha v} \to Q_{\alpha\dot\beta}$. With respect to the new rotation group the *topological supercharge*,

$$\overline{\mathcal{Q}} \equiv \epsilon^{\dot\alpha\dot\beta}\overline{Q}_{\dot\alpha\dot\beta} = \overline{Q}_{\dot1\dot2} - \overline{Q}_{\dot2\dot1} \tag{5.21}$$

is a scalar with respect to $\mathcal{K}'$:

$$[M_{\alpha\beta}, \overline{\mathcal{Q}}] = 0, \qquad [M'_{\dot\alpha\dot\beta}, \overline{\mathcal{Q}}] = 0. \tag{5.22}$$

The topological supercharge $\overline{\mathcal{Q}}$ will provide the topological symmetry δ that we need for the theory to be topological. The $\mathcal{N} = 2$ supersymmetric algebra also gives a natural way to construct the operator G_μ defined in (5.11). In fact, define

$$G_\mu = \frac{i}{4}(\overline\sigma_\mu)^{\dot\alpha\gamma} Q_{\gamma\dot\alpha}. \tag{5.23}$$

Using now the $\{Q, \overline{Q}\}$ anticommutator in (5.14) it is easy to show that

$$\{\overline{\mathcal{Q}}, G_\mu\} = \partial_\mu. \tag{5.24}$$

This means that the supersymmetry algebra by itself almost guarantees (5.2). In the models that we will consider (5.2) is true (at least on-shell). Finally, notice that from the anticommutator $\{\overline{Q}, \overline{Q}\}$ in (5.14) follows that the topological supercharge is nilpotent (in the absence of central charge):

$$\overline{\mathcal{Q}}^2 = 0. \tag{5.25}$$

The main conclusion of this section is that *by twisting $\mathcal{N} = 2$ supersymmetry one can construct a quantum field theory which satisfies (almost) all the requirements of a topological quantum field theory of the Witten type.* In the rest of this chapter we are going to analyse in detail the twisting of the $\mathcal{N} = 2$ supersymmetric theories that we described in previous chapters.

5.3. Donaldson–Witten theory

Donaldson–Witten theory (also known as topological Yang–Mills theory) is the topological quantum field theory that results from twisting $\mathcal{N} = 2$ supersymmetric Yang–Mills theory in four dimensions. Historically it was the first TQFT of the Witten type, and, as we will see, it provides a realization of Donaldson theory.

As described in section 4.4 $\mathcal{N} = 2$ supersymmetric Yang–Mills theory contains a gauge field A_μ, two spinors $\lambda_{v\alpha}$, and a complex scalar ϕ, all of them in the adjoint representation of the gauge group $\mathcal{G}$. In the off-shell formulation we also have auxiliary fields D_{vw} in the **3** of the internal $SU(2)_R$. Under the twisting the fields in the $\mathcal{N} = 2$ supersymmetric multiplet change their spin content as follows:

$$
\begin{aligned}
A_\mu \; (\mathbf{2},\mathbf{2},\mathbf{0})^0 \;&\rightarrow\; A_\mu \; (\mathbf{2},\mathbf{2})^0, \\
\lambda_{\alpha v} \; (\mathbf{2},\mathbf{0},\mathbf{2})^{-1} \;&\rightarrow\; \psi_{\alpha\dot\beta} \; (\mathbf{2},\mathbf{2})^1, \\
\bar\lambda_{\dot\alpha v} \; (\mathbf{0},\mathbf{2},\mathbf{2})^1 \;&\rightarrow\; \eta \; (\mathbf{0},\mathbf{0})^{-1}, \quad \chi_{\dot\alpha\dot\beta} \; (\mathbf{1},\mathbf{0})^{-1}, \\
\phi \; (\mathbf{0},\mathbf{0},\mathbf{0})^{-2} \;&\rightarrow\; \phi \; (\mathbf{0},\mathbf{0})^{-2}, \\
\phi^\dagger \; (\mathbf{0},\mathbf{0},\mathbf{0})^2 \;&\rightarrow\; \phi^\dagger \; (\mathbf{0},\mathbf{0})^2, \\
D_{vw} \; (\mathbf{0},\mathbf{0},\mathbf{1})^0 \;&\rightarrow\; D_{\dot\alpha\dot\beta} \; (\mathbf{0},\mathbf{0})^0,
\end{aligned}
\tag{5.26}
$$

where we have written the quantum numbers with respect to the total symmetry group $\mathcal{H}$ (5.16) before the twisting, and with respect to the group $\mathcal{H}' = SU(2)'_+ \otimes SU(2)_- \otimes U(1)_R$ after the twisting. In the topological theory the $U(1)_R$ charge is usually called the *ghost number*. The η and χ fields are given by the antisymmetric and symmetric pieces of $\bar\lambda_{\dot\alpha\dot\beta}$, respectively. More precisely

$$
\chi_{\dot\alpha\dot\beta} = \bar\lambda_{(\dot\alpha\dot\beta)}, \quad \eta = \frac{1}{2}\epsilon^{\dot\alpha\dot\beta}\bar\lambda_{\dot\alpha\dot\beta}.
\tag{5.27}
$$

From the $\mathcal{N} = 2$ supersymmetric action (4.63) it is straightforward to find (after continuation to Euclidean space) the twisted action on an arbitrary Riemmanian four-manifold endowed with a metric $g_{\mu\nu}$:

$$
\begin{aligned}
S = \frac{1}{g^2} \int d^4x \sqrt{g}\, \mathrm{Tr} \Big(&\nabla_\mu \phi \nabla^\mu \phi^\dagger - i\psi^{\dot\beta}{}_\alpha \bar\sigma^{\mu\dot\alpha\alpha} \nabla_\mu \chi_{\dot\alpha\dot\beta} - i\psi_{\alpha\dot\alpha} \nabla^{\dot\alpha\alpha}\eta - \frac{1}{4} F_{\mu\nu} F^{\mu\nu} \\
&+ \frac{1}{4} D_{\dot\alpha\dot\beta} D^{\dot\alpha\dot\beta} - \frac{1}{2}[\phi,\phi^\dagger]^2 - \frac{i}{\sqrt{2}} \chi^{\dot\alpha\dot\beta}[\phi,\chi_{\dot\alpha\dot\beta}] \\
&+ i\sqrt{2}\eta[\phi,\eta] - \frac{i}{\sqrt{2}}\psi_{\alpha\dot\alpha}[\psi^{\alpha\dot\alpha},\phi^\dagger] \Big).
\end{aligned}
\tag{5.28}
$$

The reader should not be confused between the coupling constant g and the square root of the determinant of the metric, $\sqrt{g} = (\det(g_{\mu\nu}))^{1/2}$. The $\overline{Q}$-transformations are easily obtained from (5.21) and the $\mathcal{N} = 2$ supersymmetric transformations (5.14):

$$
\begin{aligned}
[\overline{Q}, \phi] &= 0, \\
[\overline{Q}, A_\mu] &= \psi_\mu, \\
\{\overline{Q}, \eta\} &= [\phi, \phi^\dagger], \\
\{\overline{Q}, \psi_\mu\} &= 2\sqrt{2}\nabla_\mu \phi,
\end{aligned}
\qquad
\begin{aligned}
[\overline{Q}, \phi^\dagger] &= 2\sqrt{2}i\eta, \\
\{\overline{Q}, \chi_{\dot\alpha\dot\beta}\} &= i(F^+_{\dot\alpha\dot\beta} - D_{\dot\alpha\dot\beta}), \\
[\overline{Q}, D] &= (2\nabla\psi)_+ + 2\sqrt{2}[\phi, \chi].
\end{aligned}
\qquad (5.29)
$$

In (5.29) $\psi_\mu = \sigma_{\mu\alpha\dot\beta}\psi^{\alpha\dot\beta}$ and $F^+_{\dot\alpha\dot\beta} = \sigma^{\mu\nu}_{\dot\alpha\dot\beta}F_{\mu\nu}$ is the self-dual part of $F_{\mu\nu}$ (see Appendix A for more details). Notice that $\overline{Q}^2$ is a gauge transformation, in accordance with (5.5). The origin of this property comes from the twisting having been made from a supersymmetric theory in the Wess–Zumino gauge in which supersymmetric transformations close up to gauge transformations. It is not difficult to show that the action of Donaldson–Witten theory (5.28) is $\overline{Q}$-exact up to a topological term, i.e.,

$$
S = \{\overline{Q}, V\} - \frac{1}{2}\int F \wedge F, \qquad (5.30)
$$

where

$$
V = \int d^4x\,\sqrt{g}\,\mathrm{Tr}\left(\frac{i}{4}\chi_{\alpha\beta}(F^{\alpha\beta} + D^{\alpha\beta}) - \frac{1}{2}\eta[\phi, \phi^\dagger] + \frac{1}{2\sqrt{2}}\psi_{\alpha\dot\alpha}\nabla^{\dot\alpha\alpha}\phi^\dagger\right). \quad (5.31)
$$

As we will see in a moment, this has important implications for the quantum behavior of the theory.

One of the most interesting aspects of the twisting procedure is the following: if we set the original $\mathcal{N} = 2$ supersymmetric Yang–Mills theory on an arbitrary Riemannian four-manifold, using the well known prescription of minimal coupling to gravity, we find global obstructions to have a well defined theory. The reason is very simple: not every four-manifold is Spin, and therefore the fermionic fields $\lambda_{\alpha v}$ are not well defined unless $w_2(X) = 0$. However all fields are differential forms on X after the twisting, therefore the twisted theory makes sense globally on an arbitrary Riemannian four-manifold.

The observables of Donaldson–Witten theory can be constructed by using the topological descent equations. As we have emphasized, these equations have a canonical solution given by the operator G_μ in (5.23). Using again the

$\mathcal{N} = 2$ supersymmetry transformations (5.14) one can work out the action of G_μ on the different fields of the theory. The result is:

$$
\begin{aligned}
[G_\mu, \phi] &= \frac{1}{2\sqrt{2}} \psi_\mu, & [G, \overline{\phi}] &= 0, \\
[G_\nu, A_\mu] &= \frac{i}{2} g_{\mu\nu} \eta - i\chi_{\mu\nu}, & [G, F^+] &= i\nabla\chi + \frac{3i}{2} * \nabla\eta, \\
[G, \eta] &= -\frac{i\sqrt{2}}{4} \nabla\overline{\phi}, & \{G, \chi\} &= -\frac{3i\sqrt{2}}{8} * \nabla\overline{\phi}, \\
\{G_\mu, \psi_\nu\} &= -(F_{\mu\nu}^- + D_{\mu\nu}^+), & [G, D] &= -\frac{3i}{4} * \nabla\eta + \frac{3i}{2} \nabla\chi.
\end{aligned}
\tag{5.32}
$$

We can now construct the topological observables of the theory by using the descent equations. The starting point must be a set of gauge-invariant, $\overline{\mathcal{Q}}$-closed operators which are not $\overline{\mathcal{Q}}$-trivial. Since $[\overline{\mathcal{Q}}, \phi] = 0$ these operators are the gauge-invariant polynomials in the field ϕ. For a simple gauge group of rank r the algebra of these polynomials is generated by r elements, and we shall denote this basis by $\mathcal{O}_n$, $n = 1, \ldots, r$. For $SU(N)$ one can take for example

$$
\mathcal{O}_n = \text{Tr}\,(\phi^{n+1}), \quad n = 1, \ldots, N.
\tag{5.33}
$$

Starting from these operators the descent procedure produces a set of operators $\mathcal{O}_n^{(i)}$, $i = 1, \ldots, 4$. In most of this book we shall restrict ourselves to the gauge group $SU(2)$, therefore the starting point for the descent procedure will be the operator

$$
\mathcal{O} = \text{Tr}\,(\phi^2).
\tag{5.34}
$$

Using now (5.32) we can obtain the canonical solution to the descent equations. The first two topological descendants obtained in this way are

$$
\begin{aligned}
\mathcal{O}^{(1)} &= \text{Tr}\left(\frac{1}{\sqrt{2}} \phi\psi_\mu\right) dx^\mu, \\
\mathcal{O}^{(2)} &= -\frac{1}{2} \text{Tr}\left(\frac{1}{\sqrt{2}} \phi(F_{\mu\nu}^- + D_{\mu\nu}) - \frac{1}{4}\psi_\mu\psi_\nu\right) dx^\mu \wedge dx^\nu.
\end{aligned}
\tag{5.35}
$$

As we shall see, supersymmetric configurations have $D = F^+$, and in that subspace one can write $F^- + D = F$. In this book we will restrict ourselves to the observables:

$$
I_1(\delta) = \int_\delta \mathcal{O}^{(1)}, \qquad I_2(S) = \int_S \mathcal{O}^{(2)},
\tag{5.36}
$$

where $\delta \in H_1(X)$, $S \in H_2(X)$. They correspond to the differential forms on the moduli space of ASD connections which were introduced in (2.55) through the use of the Donaldson map (2.54) (and this is why we have used the same notation for both). Notice that the ghost number of the operators in (5.35) is, in fact, their degree as differential forms in moduli space. The operators (5.35) are naturally interpreted as the decomposition of the Pontrjagin class of the universal bundle (3.17) with respect to the bi-grading of $\Omega^*(\mathcal{B}^* \times X)$. For example, the Grassmannian field ψ_μ can be interpreted as a $(1, 1)$ form: a one-form in space-time and also a one-form in the space of connections $\mathcal{A}$. The operator $\overline{\mathcal{Q}}$ is then interpreted as the equivariant differential in $\mathcal{A}$ with respect to gauge transformations. This leads to a beautiful geometric interpretation of topological Yang–Mills theory in terms of equivariant cohomology. There is also a beautiful geometrical interpretation of this theory in terms of the Mathai–Quillen formalism, which will be considered in the next chapter.

Now that we have formulated the model and constructed its observables, let us consider the computation of correlation functions. Since the action of the twisted theory is $\overline{\mathcal{Q}}$-exact up to a topological term, we will consider the topological theory defined by the Donaldson–Witten action $S_{\mathrm{DW}} = \{\overline{\mathcal{Q}}, V\}$, where V is defined in (5.31). The functional integral of the theory defined by the Donaldson–Witten action can be drastically simplified by taking into account the following fact. The (un-normalized) correlation functions of the theory are defined by

$$\langle \phi_1 \cdots \phi_n \rangle = \int \mathcal{D}\phi \, \phi_1 \cdots \phi_n \, e^{-\frac{1}{g^2} S_{\mathrm{DW}}}, \qquad (5.37)$$

where $\phi_1, \ldots, \phi_n$ are generic fields which are invariant under $\overline{\mathcal{Q}}$ transformations. Since S_{DW} is $\overline{\mathcal{Q}}$-exact one has:

$$\frac{\partial}{\partial g}\langle \phi_1 \cdots \phi_n \rangle = \frac{2}{g^3}\langle \phi_1 \cdots \phi_n S_{\mathrm{DW}} \rangle = \frac{2}{g^3}\langle \{\overline{\mathcal{Q}}, \phi_1 \cdots \phi_n V\} \rangle = 0, \qquad (5.38)$$

where we have used that $\overline{\mathcal{Q}}$ is a symmetry of the theory, and therefore the insertion of a $\overline{\mathcal{Q}}$-exact operator in the functional integral gives zero. The above result is remarkable: it says that in a topological quantum field theory in which the action is $\overline{\mathcal{Q}}$-exact the computation of correlation functions of products of $\overline{\mathcal{Q}}$-invariant operators does not depend on the value of the coupling constant. In particular, *the semi-classical approximation is exact!* We can then evaluate the functional integral in the saddle-point approximation as follows: first, we look at zero-modes, *i.e.*, classical configurations that minimize the action. Then we look at non zero-modes, *i.e.*, we consider quantum

fluctuations around these configurations. Since the saddle point approximation is exact it is enough to consider the quadratic fluctuations. The integral
over the zero-modes gives a finite integral over the space of bosonic collective coordinates, and a finite Grassmannian integral over the zero-modes of
the fermionic fields. The integral over the quadratic fluctuations involves the
determinants of the kinetic operators in S_{DW}. Since the theory has a Bose–
Fermi $\overline{\mathcal{Q}}$ symmetry it turns out that the determinants cancel (up to a sign),
as in supersymmetric theories.

Let us then analyse the bosonic and fermionic zero-modes. A quick way
to find the bosonic zero-modes is to look for supersymmetric configurations.
These are classical configurations such that $\{\overline{\mathcal{Q}}, \mathrm{Fermi}\} = 0$ for all Fermi fields
in the theory, and they give minima of the Lagrangian. Indeed, it is known
that in topological quantum field theories with a fermionic symmetry $\overline{\mathcal{Q}}$ one
can compute by localization on the fixed points of this symmetry. In this
case, by looking at $\{\overline{\mathcal{Q}}, \chi\} = 0$ one finds

$$F^+ = D^+. \tag{5.39}$$

But on-shell $D^+ = 0$, and therefore (5.39) reduces to the usual ASD equations. The zero-modes of the gauge field are then instanton configurations.
In addition, by looking at $\{\overline{\mathcal{Q}}, \psi\} = 0$ we find the equation of motion for the
ϕ field,

$$\nabla \phi = 0. \tag{5.40}$$

This equation is also familiar: as we saw in Chapter 1, its non-trivial solutions
correspond to reducible connections. Let us assume for simplicity that we are
in a situation in which no reducible solutions occur, so that the only solution
to (5.40) is $\phi = 0$. In that case (5.39) tells us that the integral over the
collective coordinates reduces to an integral over the instanton moduli space
$\mathcal{M}_{\mathrm{ASD}}$.

Let us now look at the fermionic zero-modes in the background of an
instanton. The kinetic terms for the ψ, χ and η fermions fit precisely into the
instanton deformation complex (2.46). Therefore using the index theorem we
can compute

$$N_\psi - N_\chi = \dim \mathcal{M}_{\mathrm{ASD}}, \tag{5.41}$$

where $N_{\psi,\chi}$ denotes the number of zero-modes of the corresponding fields,
and we have used that the connection A is irreducible, so that η (which is

a scalar) has no zero-modes (in other words, $\nabla \eta = 0$ only has the trivial solution). Finally, if we assume that the connection is regular then one has that $\text{Coker}\,(p^+\nabla) = 0$, and there are not χ zero-modes. In this situation the number of ψ zero-modes is simply the dimension of the moduli space of ASD instantons. If we denote the bosonic and the fermionic zero-modes by da_i, $d\psi_i$, respectively, where $i = 1, \ldots, D$ and $D = \dim \mathcal{M}_{\text{ASD}}$, then the zero-mode measure becomes:

$$\prod_{i=1}^{D} da_i \, d\psi_i. \tag{5.42}$$

This is, in fact, the natural measure for integration of differential forms on $\mathcal{M}_{\text{ASD}}$, and the Grassmannian variables ψ_i are then interpreted as a basis of one-forms on $\mathcal{M}_{\text{ASD}}$.

We can already discuss how to compute correlation functions of the operators $\mathcal{O}$, $I_2(S)$, and $I_1(\delta)$. These operators contain the fields ψ, A_μ and ϕ. In evaluating the functional integral it is enough to replace ψ and A_μ by their zero-modes, and the field ϕ (with no zero-modes) by its quantum fluctuations, which we then integrate out at quadratic order. Further corrections which are higher order in the coupling constant do not contribute to the saddle point approximation, which in this case is exact. We then have to compute the one-point correlation function $\langle \phi^a \rangle$. The relevant terms in the action are

$$S(\phi, \phi^\dagger) = \int d^4x \, \text{Tr}\Big(\nabla_\mu \phi \nabla^\mu \phi^\dagger - \frac{i}{\sqrt{2}} \phi^\dagger [\psi_\mu, \psi^\mu]\Big), \tag{5.43}$$

since we are only considering quadratic terms. We then have to compute

$$\langle \phi^a(x) \rangle = \int \mathcal{D}\phi \mathcal{D}\phi^\dagger \phi^a(x) \exp\Big(-S(\phi, \phi^\dagger)\Big). \tag{5.44}$$

If we take into account that

$$\langle \phi^a(x) \phi^{b\dagger}(y) \rangle = G^{ab}(x - y), \tag{5.45}$$

where $G^{ab}(x - y)$ is the Green's function of the Laplacian $\nabla_\mu \nabla^\mu$, we find

$$\langle \phi^a(x) \rangle = -\frac{i}{\sqrt{2}} \int d^4y \sqrt{g}\, G^{ab}(x, y) [\psi(x)_\mu, \psi(y)^\mu]_b. \tag{5.46}$$

This expresses ϕ in terms of zero-modes. It turns out that this is precisely (up to a constant) the component along $\mathcal{B}^*$ of the curvature $K_{\mathbf{P}}$ of the universal bundle. This is in perfect agreement with the correspondence between the observables (5.35) and the differential forms on moduli space (2.55) constructed in Donaldson theory.

The main conclusion of this analysis is that, up to normalizations,

$$
\begin{aligned}
&\langle \mathcal{O}^\ell I_2(S_{i_1}) \cdots I_2(S_{i_p}) I_1(\delta_{j_1}) \cdots I_1(\delta_{j_q}) \rangle \\
&= \int_{\mathcal{M}_{\mathrm{ASD}}} \mathcal{O}^\ell \wedge I_2(S_{i_1}) \wedge \cdots \wedge I_2(S_{i_p}) \wedge I_1(\delta_{j_1}) \wedge \cdots \wedge I_1(\delta_{j_q}),
\end{aligned}
\tag{5.47}
$$

i.e., the correlation function of the observables of twisted $\mathcal{N} = 2$ supersymmetric Yang–Mills theory is precisely the corresponding Donaldson invariant (2.57). The requirement that the differential form in the right hand side has top degree (otherwise the invariant is zero) corresponds, in the field theory side, to the requirement that the correlation function has its ghost number equal to $\dim \mathcal{M}_{\mathrm{ASD}}$, *i.e.*, that the operator in the correlation function soaks up all the fermionic zero-modes, which is the well known 't Hooft rule.

Equation (5.47) establishes the equivalence between twisted $\mathcal{N} = 2$ supersymmetric Yang–Mills theory and Donaldson theory. In terms of the generating functional (2.61) we have that

$$
Z_{\mathrm{DW}}^w(p, S, \delta) = \Big\langle \exp\big(p\mathcal{O} + I_1(\delta) + I_2(S)\big) \Big\rangle_w,
\tag{5.48}
$$

where the subscript w means that we are computing the correlation functions at fixed Stiefel–Whitney class w, and as in (2.61) we are summing over all instanton numbers.

This equivalence between Donaldson theory and twisted $\mathcal{N} = 2$ supersymmetric theory opens a completely different approach to studying Donaldson invariants by means of topological quantum field theory. As will be shown in the following chapters, topological quantum field theory provides a recipe to compute the correlation functions (5.48). This result opens an entirely new way to understanding Donaldson invariants, establishing a connection between these invariants and Seiberg–Witten invariants.

As a final remark we would like to note that many features of the evaluation of the functional integral of Donaldson–Witten theory developed in this section —such as the exactness of the semi-classical approximation— have simple analogs in the evaluation of finite-dimensional integrals, and are common to most topological field theories of the cohomological type. Also typical from cohomological theories is the reduction of functional integrals to finite-dimensional integrals involving only the zero-modes. These features will become clearer in the next chapter when we develop the Mathai–Quillen formalism.

5.4. Twisted $\mathcal{N} = 2$ supersymmetric hypermultiplet

In this section we are going to analyse the twisting of the free $\mathcal{N} = 2$ supersymmetric hypermultiplet considered in section 5.5. The field content of the hypermultiplet (4.71) is given by a doublet of complex scalar fields, q_v, two Weyl fermions ψ y χ, and a doublet of complex auxiliaries F_v. After the twisting the fields in the theory become:

$$
\begin{aligned}
q_v\,(\mathbf{0,0,2}) &\longrightarrow M_{\dot\alpha}\,(\mathbf{2,0}), & q^{\dagger v}\,(\mathbf{0,0,2}) &\longrightarrow \overline{M}^{\dot\alpha}\,(\mathbf{2,0}), \\
\psi_\alpha\,(\mathbf{2,0,0}) &\longrightarrow \nu_\alpha\,(\mathbf{2,0}), & \overline\psi_{\dot\alpha}\,(\mathbf{0,2,0}) &\longrightarrow \overline\mu_{\dot\alpha}\,(\mathbf{0,2}), \\
\overline\chi_{\dot\alpha}\,(\mathbf{0,2,0}) &\longrightarrow \mu_{\dot\alpha}\,(\mathbf{0,2}), & \chi_\alpha\,(\mathbf{2,0,0}) &\longrightarrow \overline\nu_\alpha\,(\mathbf{2,0}), \\
F_v\,(\mathbf{0,0,2}) &\longrightarrow K_{\dot\alpha}\,(\mathbf{2,0}), & F^{\dagger v}\,(\mathbf{0,0,2}) &\longrightarrow \overline{K}^{\dot\alpha}\,(\mathbf{2,0}),
\end{aligned}
\tag{5.49}
$$

where we have written the quantum numbers with respect to $\mathcal{H}$ and $\mathcal{H}'$. After the twisting the free action (4.73) becomes:

$$
\begin{aligned}
S = \int d^4 x\,(&\partial_\mu \overline{M}^{\dot\alpha}\partial^\mu M_{\dot\alpha} + m\overline{M}^{\dot\alpha}K_{\dot\alpha} + m\overline{K}^{\dot\alpha}M_{\dot\alpha} + \overline{K}^{\dot\alpha}K_{\dot\alpha} \\
&- i\overline\nu_\alpha \partial^{\dot\alpha\alpha}\mu_{\dot\alpha} - i\overline\mu^{\dot\alpha}\partial_{\alpha\dot\alpha}\nu^\alpha - m\overline\mu^{\dot\alpha}\mu_{\dot\alpha} - m\overline\nu_\alpha\nu^\alpha),
\end{aligned}
\tag{5.50}
$$

whilst the topological transformations, after using (4.75), turn out to be:

$$
\begin{aligned}
\delta M_{\dot\alpha} &= -\sqrt{2}\epsilon\mu_{\dot\alpha}, & \delta \overline{M}_{\dot\alpha} &= \sqrt{2}\epsilon\overline\mu_{\dot\alpha}, \\
\delta \mu_{\dot\alpha} &= -\sqrt{2}\epsilon K_{\dot\alpha}, & \delta \overline\mu_{\dot\alpha} &= -\sqrt{2}\epsilon\overline{K}_{\dot\alpha}, \\
\delta \nu_\alpha &= i\sqrt{2}\epsilon\partial_{\alpha\dot\alpha}M^{\dot\alpha}, & \delta \overline\nu_\alpha &= i\sqrt{2}\epsilon\partial_{\alpha\dot\alpha}\overline{M}^{\dot\alpha}, \\
\delta K_{\dot\alpha} &= -i\sqrt{2}\epsilon\partial_{\alpha\dot\alpha}\nu^\alpha, & \delta \overline{K}_{\dot\alpha} &= i\sqrt{2}\epsilon\partial_{\alpha\dot\alpha}\overline\nu^\alpha.
\end{aligned}
\tag{5.51}
$$

This off-shell formulation of the theory is not particularly suitable since the action turns out not to be δ-exact. However, there exists another off-shell formulation which cures this problem, it requires the introduction of new auxiliary fields h y $\overline{h}$. We introduce as well new δ' transformations:

$$
\begin{aligned}
\delta' M_{\dot\alpha} &= -\sqrt{2}\epsilon\mu_{\dot\alpha}, & \delta'\overline{M}_{\dot\alpha} &= \sqrt{2}\epsilon\overline\mu_{\dot\alpha}, \\
\delta' \mu_{\dot\alpha} &= -\sqrt{2}\epsilon m M_{\dot\alpha}, & \delta'\overline\mu_{\dot\alpha} &= -\sqrt{2}\epsilon m\overline{M}_{\dot\alpha}, \\
\delta' \nu_\alpha &= \sqrt{2}\epsilon h_\alpha, & \delta'\overline\nu_\alpha &= \sqrt{2}\epsilon\overline{h}_\alpha, \\
\delta' h_\alpha &= \sqrt{2}\epsilon m\nu_\alpha, & \delta'\overline{h}_\alpha &= -\sqrt{2}\epsilon m\overline\nu_\alpha.
\end{aligned}
\tag{5.52}
$$

In this formulation the central charge is realized off-shell as a global $U(1)$ symmetry with parameter m, and the action can be written as $\epsilon \delta' \Lambda$, where,

$$
\Lambda = \int d^4x \, \frac{1}{\sqrt{2}} \Big(-i\overline{\nu}_\alpha \partial^{\dot\alpha\alpha} M_{\dot\alpha} - i\overline{M}^{\dot\alpha} \partial_{\alpha\dot\alpha} \nu^\alpha - \frac{1}{2} m\overline{\mu}^{\dot\alpha} M_{\dot\alpha}
$$
$$
- \frac{1}{2}\overline{\nu}_\alpha h^\alpha - \frac{1}{2} m\overline{M}^{\dot\alpha} \mu_{\dot\alpha} - \frac{1}{2}\overline{h}_\alpha \nu^\alpha \Big),
$$

(5.53)

and reads,

$$
S = \int d^4x \, \Big(-i\overline{h}_\alpha \partial^{\alpha\alpha} M_{\dot\alpha} - i\overline{M}^{\dot\alpha} \partial_{\alpha\dot\alpha} h^\alpha - i\overline{\nu}_\alpha \partial^{\dot\alpha\alpha} \mu_{\dot\alpha} - i\overline{\mu}^{\dot\alpha} \partial_{\alpha\dot\alpha} \nu^\alpha
$$
$$
- m^2 \overline{M}^{\dot\alpha} M_{\dot\alpha} - m\overline{\nu}_\alpha \nu^\alpha - m\overline{\mu}^{\dot\alpha} \mu_{\dot\alpha} - \overline{h}_\alpha h^\alpha \Big).
$$

(5.54)

After integrating the new auxiliary fields in this action, and the ones in (5.50), one finds both actions are equivalent on-shell.

Notice that, in contrast to $\mathcal{N} = 2$ supersymmetric Yang–Mills, where the twisting procedure made the theory well defined on any four-manifold, the twisted free hypermultiplet theory contains spinors. In fact, the twisting has converted the two scalar fields q_v into a new spinor field $M_{\dot\alpha}$. Strictly speaking, the twisted free hypermultiplet only makes sense globally on Spin manifolds. We will see later that under certain circumstances the coupling to gauge fields can solve this problem.

5.5. Extensions of Donaldson–Witten theory

We can now put together the results obtained for the twisted $\mathcal{N} = 2$ supersymmetric Yang–Mills theory and for the free hypermultiplet and consider the coupled theory. Again, it is convenient to redefine the auxiliary fields for the hypermultiplet in the way that we have just described. The results are easily derived following the same procedure as in the previous sections. For the action we find:

$$S = \int d^4x \, \mathrm{Tr} \left(\nabla_\mu \phi \nabla^\mu \phi^\dagger + i\psi^{\dot\beta}{}_\alpha \nabla^{\dot\alpha\alpha} \chi_{\dot\alpha\dot\beta} + i\psi_{\alpha\dot\alpha} \nabla^{\dot\alpha\alpha}\eta + \frac{1}{4} F_{\mu\nu} F^{\mu\nu} \right.$$

$$- \frac{1}{4} D_{\dot\alpha\dot\beta} D^{\dot\alpha\dot\beta} + \frac{1}{2} [\phi,\phi^\dagger]^2 - \frac{i}{\sqrt{2}} \chi^{\dot\alpha\dot\beta}[\phi,\chi_{\dot\alpha\dot\beta}] - i\sqrt{2}\eta[\phi,\eta]$$

$$\left. - \frac{i}{\sqrt{2}} \psi_{\alpha\dot\alpha}[\psi^{\alpha\dot\alpha},\phi^\dagger] \right)$$

$$+ \int d^4x \left(-i\bar{h}_\alpha \nabla^{\alpha\dot\alpha} M_{\dot\alpha} - i\overline{M}^{\dot\alpha} \nabla_{\alpha\dot\alpha} h^\alpha - i\bar{\nu}_\alpha \nabla^{\dot\alpha\alpha}\mu_{\dot\alpha} - i\bar{\mu}^{\dot\alpha} \nabla_{\alpha\dot\alpha}\nu^\alpha \right.$$

$$\left. - m^2 \overline{M}^{\dot\alpha} M_{\dot\alpha} - m\bar{\nu}_\alpha \nu^\alpha - m\bar{\mu}^{\dot\alpha}\mu_{\dot\alpha} - \bar{h}_\alpha h^\alpha - \overline{M}^{\dot\alpha} F^+_{\dot\alpha\dot\beta} M^{\dot\beta} \right)$$

$$+ \int d^4x \left(i\sqrt{2}(\overline{M}^{\dot\alpha} \chi_{\dot\alpha}{}^{\dot\beta} \mu_{\dot\beta} + \bar{\mu}^{\dot\alpha} \chi_{\dot\alpha}{}^{\dot\beta} M_{\dot\beta}) + i\sqrt{2}(\overline{M}^{\dot\alpha} \psi_{\alpha\dot\alpha}\nu^\alpha - \bar{\nu}_\alpha \psi^{\dot\alpha\alpha} M_{\dot\alpha}) \right.$$

$$+ \overline{M}^{\dot\alpha}\{\phi,\phi^\dagger\}M_{\dot\alpha} + i\sqrt{2}(\overline{M}^{\dot\alpha}\eta\mu_{\dot\alpha} - \bar{\mu}^{\dot\alpha}\eta M_{\dot\alpha}) - i\sqrt{2}\bar{\nu}_\alpha \phi\nu^\alpha$$

$$+ i\sqrt{2}\bar{\mu}^{\dot\alpha}\phi^\dagger \mu_{\dot\alpha} + (\overline{M}^{(\dot\alpha} T^a M^{\dot\beta)})(\overline{M}_{(\dot\alpha} T^a M_{\dot\beta)}) - i\sqrt{2}m\overline{M}^{\dot\alpha}\phi M_{\dot\alpha}$$

$$\left. + i\sqrt{2}m\overline{M}^{\dot\alpha}\phi^\dagger M_{\dot\alpha} \right),$$

$$(5.55)$$

and the δ' transformations become:

$$\delta'\phi = 0, \qquad \delta'\phi^\dagger = 2\sqrt{2}i\epsilon\eta, \qquad \delta'\eta = i\epsilon[\phi,\phi^\dagger],$$

$$\delta'A_\mu = \epsilon\psi_\mu, \qquad \delta'\psi_\mu = 2\sqrt{2}\epsilon\nabla_\mu\phi,$$

$$\delta'\chi^a{}_{\alpha\beta} = i\epsilon(F^{+a}_{\dot\alpha\dot\beta} - D^a_{\dot\alpha\dot\beta} + 2i\overline{M}_{(\dot\alpha} T^a M_{\dot\beta)}),$$

$$\delta'D^a_{\dot\alpha\dot\beta} = \epsilon(d_A\psi)^a_{\dot\alpha\dot\beta} + 2\sqrt{2}i\epsilon[\chi^{\dot\alpha\dot\beta},\phi]^a + 2\sqrt{2}\epsilon(\overline{M}^{(\dot\alpha} T^a \mu^{\dot\beta)} + \bar{\mu}^{(\dot\alpha} T^a M^{\dot\beta)}),$$

$$\delta'M_{\dot\alpha} = -\sqrt{2}\epsilon\mu_{\dot\alpha}, \qquad \delta'\overline{M}_{\dot\alpha} = \sqrt{2}\epsilon\bar{\mu}_{\dot\alpha},$$

$$\delta'\mu^j{}_{\dot\alpha} = \sqrt{2}\epsilon m M^j{}_{\dot\alpha} + 2i\epsilon\phi^{jk} M^k{}_{\dot\alpha}, \qquad \delta'\bar{\mu}^j{}_{\dot\alpha} = \sqrt{2}\epsilon m\overline{M}^j{}_{\dot\alpha} + 2i\epsilon\overline{M}^k{}_{\dot\alpha}\phi^{kj},$$

$$\delta'\nu_\alpha = \sqrt{2}\epsilon h_\alpha, \qquad \delta'\bar{\nu}_\alpha = \sqrt{2}\epsilon\bar{h}_\alpha,$$

$$\delta'h^j{}_\alpha = -\sqrt{2}\epsilon m\nu^j{}_\alpha - 2i\epsilon\phi^{jk}\nu^k{}_\alpha, \qquad \delta'\bar{h}^j{}_\alpha = \sqrt{2}\epsilon m\bar{\nu}^j{}_\alpha + 2i\epsilon\bar{\nu}^k{}_\alpha\phi^{kj}.$$

$$(5.56)$$

Under these transformations (5.55) is δ'-exact,

$$\epsilon S = \delta' \int d^4x \, \mathrm{Tr} \left(\frac{i}{4}\chi_{\dot\alpha\dot\beta} F^{\dot\alpha\dot\beta} + \frac{1}{4}\chi_{\dot\alpha\dot\beta} D^{\dot\alpha\dot\beta} - i\eta[\phi,\phi^\dagger] + \frac{i}{\sqrt{2}}\psi_{\alpha\dot\alpha}\nabla^{\dot\alpha\alpha}\eta \right)$$

$$+ \int d^4x \, \frac{1}{\sqrt{2}} \left(-i\bar{\nu}_\alpha \nabla^{\dot\alpha\alpha} M_{\dot\alpha} - i\overline{M}^{\dot\alpha}\nabla_{\alpha\dot\alpha}\nu^\alpha - \frac{1}{2}m\bar{\mu}^{\dot\alpha} M_{\dot\alpha} - \frac{1}{2}\bar{\nu}_\alpha h^\alpha \right.$$

$$- \frac{1}{2}m\overline{M}^{\dot\alpha}\mu_{\dot\alpha} - \frac{1}{2}\bar{h}_\alpha\nu^\alpha + \frac{i}{\sqrt{2}}(\bar{\mu}^{\dot\alpha}\phi^\dagger M_{\dot\alpha} + \overline{M}^{\dot\alpha}\phi^\dagger\mu_{\dot\alpha})$$

$$\left. - \frac{i}{\sqrt{2}}\overline{M}_{\dot\alpha}\chi^{\dot\alpha\dot\beta} M_{\dot\beta} \right),$$

$$(5.57)$$

up to a topological term $\sim \int F \wedge F$. The observables of this theory are the same as in Donaldson–Witten theory, in other words there are no new observables associated with the hypermultiplet fields.

5.6. Monopole equations

Let us consider the twisted $\mathcal{N} = 2$ supersymmetric theory coupled to a hypermultiplet that we have just analysed, but in the simple case of an abelian group $U(1)$ and with zero mass, $m = 0$. Since the action is $\overline{Q}$-exact the theory can be analysed very much like Donaldson–Witten theory. In order to compute correlation functions we can restrict ourselves to supersymmetric configurations. The auxiliary fields D and h satisfy the algebraic equations of motion $D_{\dot\alpha\dot\beta} = 0$ and $h_\alpha = \nabla_{\alpha\dot\alpha} M^{\dot\alpha}$. Therefore $\{\overline{Q}, \chi\} = 0$ gives

$$F^+_{\dot\alpha\dot\beta} + 2i\overline{M}_{(\dot\alpha} M_{\dot\beta)} = 0, \qquad (5.58)$$

whilst $\{\overline{Q}, \nu_\alpha\} = 0$ gives

$$\nabla_{\alpha\dot\alpha} M^{\dot\alpha} = 0. \qquad (5.59)$$

These are precisely the Seiberg–Witten equations (3.2). The only subtlety is, of course, that in the original equations (3.2) M is a section of $S^+ \otimes L^{1/2}$, where L is the determinant line bundle of a Spin_c structure, whilst here M is a section of $S^+ \otimes U$, where U is in principle a conventional $U(1)$ bundle. In fact, as we remarked before, the twisted hypermultiplet is not globally well defined on an arbitrary four-manifold, unless $w_2(X) = 0$. As we shall see, in the relevant physical realization of the $U(1)$ theory a subtle topological effect will make U the square root of the determinant line bundle L associated with a Spin_c connection, in such a way that the twisted abelian theory will be well defined. In any case, if we assume that the $U(1)$ connection is really a Spin_c connection, we see that supersymmetric configurations are given by solutions of the Seiberg–Witten equations, and the space of bosonic zero-modes is precisely the Seiberg–Witten moduli space. We have assumed here, as in our discussion of Donaldson–Witten theory, that there are no reducible pairs, and therefore $\phi = 0$. The analysis of the fermionic coordinates is also very similar to the analysis of Donaldson theory. The fermionic zero-modes fit into the deformation complex (3.10), and in a generic situation their number equals the dimension of the Seiberg–Witten moduli space.

Let us now analyse the observables of the abelian theory. The starting operator for the descent procedure is

$$\mathcal{O} = \phi, \tag{5.60}$$

and after using (5.23) one finds that the only non-trivial observables are ϕ itself and

$$\nu_i = \int_{\delta_i} \mathcal{O}^{(1)}, \quad i = 1, \dots, b_1, \tag{5.61}$$

where

$$\mathcal{O}^{(1)} = \frac{1}{2\sqrt{2}} \psi. \tag{5.62}$$

The observables ϕ and ν_i, $i = 1, \dots, b_1$, are precisely the differential forms on moduli space (3.19) and (3.20), respectively. As in Donaldson–Witten theory, it is straightforward to show that correlation functions in this topological quantum field theory are precisely the Seiberg–Witten invariants (3.22) (assuming, of course, that the $U(1)$ connection is really a Spin_c connection). More precisely, one has

$$\langle \phi^\ell \nu_{i_1} \dots \nu_{i_r} \rangle = \int_{\mathcal{M}_{\mathrm{SW}}} \phi^\ell \wedge \nu_{i_1} \wedge \dots \wedge \nu_{i_r}. \tag{5.63}$$

It is understood here that ϕ is obtained by integrating out, as in (5.46) for Donaldson–Witten theory, and is therefore expressed in terms of fermionic zero-modes.

One can also consider the moduli problem associated with a twisted Yang–Mills theory with a non-abelian gauge group G and N_f hypermultiplets in different representations of the gauge group $R_1, \dots, R_{N_f}$. The kind of moduli equations which one obtains for these models are non-abelian monopole equations. For example, in the case in which one has just one hypermultiplet these are equations for a non-abelian gauge connection A and for a spinor $M_{\dot{\alpha}}^i$ taking values in a twisted spinor bundle $S^+ \otimes E$, where E is a vector bundle associated to a principal G-bundle via a representation R of the gauge group. The equations for the moduli problem can be read as before from (5.56), and turn out to be,

$$F_{\dot\alpha\dot\beta}^{+a} + 4i\overline{M}_{(\dot\alpha}(T^a)M_{\dot\beta)} = 0,$$
$$(\nabla_E^{\alpha\dot\alpha}\overline{M}_{\dot\alpha}) = 0, \tag{5.64}$$

where T^a are the generators of the Lie algebra in the representation R, $F_{\dot\alpha\dot\beta} = F_{\dot\alpha\dot\beta}^{+a}(T^a)^{ij}$ and $\overline{M}_{(\dot\alpha}(T^a)M_{\dot\beta)}$ is a shortened form for $\overline{M}_{(\dot\alpha}^i(T^a)^{ij}M_{\dot\beta)}^j$, In (5.64) $\nabla_E^{\alpha\dot\alpha}$ is the Dirac operator for the twisted spinor bundle $S^+ \otimes E$.

The above equations only make sense in principle for a Spin manifold, since they involve spinors. There are two ways to define the theory on a general four-manifold. The most obvious one is to introduce a background Spin_c connection through an extended twisting, in such a way that the spinors M_α become sections of a bundle $S^+ \otimes L^{\frac{1}{2}} \otimes E$, where L is the determinant line bundle of the Spin_c structure. When the gauge group G is $SO(3)$ and the hypermultiplets are in the fundamental representation of the $SO(3)$ bundle V there is another possibility for defining the theory on any four-manifold X. If $w_2(V) \neq 0$ we can not lift V to an $SU(2)$ bundle E. This bundle exists locally, but there is an obstruction to defining it globally measured by $w_2(V)$. In the same way, the spinor bundle S^+ does exist locally, but globally there is an obstruction to define it measured by $w_2(X)$. However, if $w_2(V) = w_2(X)$ the bundle $S^+ \otimes E$ *does* exist, since the obstructions cancel each other. We saw a similar phenomenon in the construction of Spin_c structures, where S^+ and $L^{\frac{1}{2}}$ do not exist separately but $S^+ \otimes L^{\frac{1}{2}}$ does exist. In the split case (2.33) it is easy to check that $S^+ \otimes E$ is a well defined bundle: if $E = T^{\frac{1}{2}} \oplus T^{-\frac{1}{2}}$ it follows from (2.37) that $c_1(T^{\pm 1}) \simeq w_2(X)$, and $S^+ \otimes E$ is the sum of two Spin_c bundles.

The moduli problem for non-abelian monopoles can be analysed following the same ideas that we used for the Donaldson and Seiberg–Witten moduli problems, and we shall not repeat this analysis here, referring the reader to the references at the end of this chapter for further information.

Bibliographical notes

- Topological quantum field theories were first introduced by Witten in [43]. In that paper Donaldson–Witten theory is constructed from the twisting of $\mathcal{N} = 2$ supersymmetric Yang–Mills theory. Reviews of topological quantum field theory include for example [44][31] and [9]. In this book we shall only deal with topological quantum field theories of Witten type. Theories of Schwarz type were first considered in [45], and the most important example, Chern–Simons theory was introduced by Witten in [46]. Topological sigma models as twisted $\mathcal{N} = 2$ supersymmetric sigma models in two dimensions are introduced in [47].

- The deep relation between supersymmetry and topology was first noticed by Witten in [48] and [49], and in this last paper he formulated (and refined) Morse theory in the language of supersymmetric quantum mechanics.

- Some useful observations about topological Yang–Mills theory can be found in [50]. The uses of equivariant cohomology in this theory were first pointed out in [51], and for a detailed presentation we refer to [9].

- Twisted hypermultiplets in four dimensions (and their coupling to topological Yang–Mills fields) were considered in [52], [53], and [54] before the introduction of the Seiberg–Witten equations. After Witten recognized in [22] the importance of twisted $U(1)$ hypermultiplets for the study of four-manifold invariants, there was a renewed interest in twisted theories with hypermultiplets, which were further studied in [55][56][57][58][59].

- To make sense of non-abelian monopole equations on any four-manifold, [57] and [60] advocated the introduction of Spin$_c$ structures, whilst [61] pointed out that in the $SO(3)$ case the choice $w_2(V) = w_2(X)$ in the $SO(3)$ leads to a consistent model. Mathematically, non-abelian monopole equations have been studied in [62] and [63], as well as in [64].

Chapter 6

The Mathai–Quillen Formalism

In this chapter we shall present an alternative way of constructing the topological quantum field theories of the previous chapter based on the Mathai–Quillen formalism. This formalism applies to other cohomological theories, and we will present it in some generality, applying it to low-dimensional situations as well. This chapter can be omitted in a first reading of the book since its content can be considered as complementary. The formulation of topological quantum field theories in the Mathai–Quillen formalism is important because it provides a useful geometrical framework, but the true breakthroughs in the understanding of these theories have been provided by their formulation as twisted $\mathcal{N} = 2$ supersymmetric theories.

In general terms, topological quantum field theories of cohomological type are associated with a moduli problem characterized by three basic data: fields, symmetries, and equations. The starting point is a configuration space $\mathcal{X}$, whose elements are fields ϕ_i defined on some Riemannian manifold X. These fields are generally acted on by some group $\mathcal{G}$ of local transformations (gauge symmetries, or a diffeomorphism group, amongst others), so that one is naturally led to consider the quotient space $\mathcal{X}/\mathcal{G}$. Within this quotient space a certain subset or *moduli space*, $\mathcal{M}$, is singled out by a set of equations $s(\phi_i) = 0$:

$$\mathcal{M} = \{\phi_i \in \mathcal{X} | s(\phi_i) = 0\}/\mathcal{G}. \tag{6.1}$$

Within this framework the topological symmetry δ which we introduced in the previous chapter furnishes a representation of the $\mathcal{G}$-equivariant cohomology on the field space. When $\mathcal{G}$ is the trivial group δ is nothing but the de Rham operator on the field space. The equations $s(\phi_i)$ have to be regarded as sections of a certain vector bundle $\mathcal{E}$ over $\mathcal{X}/\mathcal{G}$. For example, in the case of Donaldson–Witten theory the configuration space $\mathcal{X}$ is the space of G-connections $\mathcal{A}$ on a bundle E over a four-manifold X, the group $\mathcal{G}$ is the group of gauge transformations associated with G, and the self-duality equations can be regarded as sections of a $\mathcal{G}$-bundle $\mathcal{E} = \mathcal{A} \times_{\mathcal{G}} \Omega^{2,+}(X, \mathbf{g}_E)$.

In ordinary finite-dimensional situations one can write the Euler class of a bundle in terms of a section and an explicit differential form representative constructed by Mathai and Quillen. In general the moduli problem of a topological quantum field theory involves an infinite-dimensional bundle $\mathcal{E}$ and a particular section s. However, it was noticed by Atiyah and Jeffrey that if one constructs the Euler class of $\mathcal{E}$ by using the section s and a natural extension of the Mathai–Quillen representative to the infinite-dimensional setting, one recovers the action principle of the theory. In particular, they showed that the partition function of the theory can be regarded as a regularized version of the Euler characteristic of $\mathcal{E}$. Conversely, the Mathai–Quillen formalism makes possible to construct a cohomological field theory starting from a moduli problem, and provides a precise geometric interpretation of all the ingredients involved in the theory.

In this chapter we first set up the Mathai–Quillen formalism in the finite-dimensional case, we illustrate it with a simple example, and then we extend it to the infinite-dimensional case. Finally, we focus on the theories of interest in this book: Donaldson–Witten theory and its extensions.

6.1. Equivariant cohomology

The Mathai–Quillen formalism is better formulated in the context of equivariant cohomology, which we briefly review here. Detailed expositions can be found in the references at the end of the chapter.

Equivariant cohomology appears when one studies a topological space X with the action of a group G. It can be realized in many different ways, but the most useful ones for us involve appropriate extensions of the de Rham cohomology. From a more general point of view, equivariant cohomology is defined as the ordinary cohomology of the space

$$X_G = EG \times_G X, \tag{6.2}$$

where EG is, as usual, the universal G-bundle. We will instead work with algebraic models, the Weil and the Cartan model. Although they are equivalent, the Cartan model is simpler and more natural in topological field theories with a gauge symmetry. However, when dealing with group actions associated with principal bundles the Cartan model must be supplemented with a horizontal projection. In these cases it is important to keep in mind the underlying Weil model.

We first formulate the Weil model of equivariant cohomology. Let G be a Lie group with Lie algebra $\mathbf{g}$. The *Weil algebra* $\mathcal{W}(\mathbf{g})$ is defined as

$$\mathcal{W}(\mathbf{g}) = \Lambda \mathbf{g}^* \otimes S \mathbf{g}^*, \tag{6.3}$$

given by the tensor product of the exterior algebra and the symmetric algebra of the dual $\mathbf{g}^*$. We take generators of $\Lambda \mathbf{g}^*$, $\{\theta^A\}_{A=1,\dots,\dim(G)}$ with degree 1, and of $S \mathbf{g}^*$, $\{u^A\}_{A=1,\dots,\dim(G)}$ with degree 2. In this way $\mathcal{W}(\mathbf{g})$ becomes a graded algebra. Let c^A_{BC} be the structure constants of $\mathbf{g}$ associated to the generators θ^A. We define a differential operator $d_{\mathcal{W}}$ from their action on the generators:

$$\begin{aligned}
d_{\mathcal{W}} \theta^A &= -\frac{1}{2} c^A_{BC} \theta^B \theta^C + u^A, \\
d_{\mathcal{W}} u^A &= c^A_{BC} u^B \theta^C.
\end{aligned} \tag{6.4}$$

We also define an inner product operator i_A as follows:

$$i_A \theta^B = \delta_{AB}, \qquad i_A u^B = 0. \tag{6.5}$$

We can also define a Lie derivative operator using the basic identity $\mathcal{L}_A = i_A d_{\mathcal{W}} + d_{\mathcal{W}} i_A$. The motivation to define such a complex comes from the following: if P is a principal G-bundle with connection $\theta \in \Omega^1(P, \mathbf{g})$ and curvature $K \in \Omega^2(P, \mathbf{g})$, we can expand θ and K as follows:

$$\theta = \theta^A T_A, \quad K = u^A T_A, \tag{6.6}$$

where $\{T_A\}_{A=1,\dots,\dim(G)}$ is a basis of $\mathbf{g}$, and $\theta^A \in \Omega^1(P)$, $u^A \in \Omega^2(P)$. Consider the inner product $\iota(C(T_A))$, where $C(T_A)$ is the fundamental vector field on P associated to the generator of the Lie algebra $\mathbf{g}$. The components defined in (6.6) verify precisely the equations (6.4) and (6.5), where the inner product i_A is understood as the geometric inner product $\iota(C(T_A))$ and the differential operator $d_{\mathcal{W}}$ is the usual de Rham operator on P. We then see that the Weil model is essentially an algebraic or universal realization of the basic relations defining connections and curvatures on principal bundles.

Consider now a G-manifold X and the complex

$$\mathcal{W}(\mathbf{g}) \otimes \Omega^*(X), \tag{6.7}$$

where $\Omega^*(X)$ is the complex of differential forms on X. On the complex (6.7) we can define a differential operator, an inner product and a Lie derivative operator by taking the tensor product of the corresponding operators in the two complexes. An element of this complex is called *basic* if it is in the kernel of the inner product operators (it is horizontal) and in the kernel of the Lie derivative operators (it is invariant). The subalgebra consisting of these elements is denoted by $\Omega_G(X)$, and its cohomology is called the *algebraic equivariant cohomology of X in the Weil model*:

$$H_G^*(X) = H^*(\Omega_G(X)). \tag{6.8}$$

As we shall see, the Weil model is specially useful for studying the cohomology of associated vector bundles and for this reason has an important role in the Mathai–Quillen formalism.

Let us now consider the Cartan model of equivariant cohomology. The fact that the construction of the Weil model involves only a subcomplex of $\mathcal{W}(\mathbf{g}) \otimes \Omega^*(X)$ suggests that a smaller complex can be chosen from the very beginning. In the Cartan model one starts with

$$S\mathbf{g}^* \otimes \Omega^*(X) \tag{6.9}$$

and defines an operator

$$d_C u^A = 0, \qquad d_C \omega = d\omega - u^A \iota(C(T_A))\omega, \tag{6.10}$$

where $\omega \in \Omega^*(X)$ and summation over A is understood. In general d_C is not nilpotent. Rather we have

$$d_C^2 = -u^A \mathcal{L}(C(T_A)), \tag{6.11}$$

but we can restrict ourselves to the invariant subcomplex

$$\Omega_G(X) = (S\mathbf{g}^* \otimes \Omega^*(X))^G, \tag{6.12}$$

where $d_C^2 = 0$. The elements in this complex are called *equivariant differential forms*. We used the same notation for the basic subcomplex in the Weil model, because in fact they are isomorphic. As $d_C^2 = 0$ on $\Omega_G(X)$, we can define the cohomology of d_C, which is precisely the *algebraic equivariant cohomology of X in the Cartan model*:

$$H_G^*(X) = H^*((S\mathbf{g}^* \otimes \Omega^*(X))^G). \tag{6.13}$$

Of course, the Weil and the Cartan models give isomorphic cohomology theories.

6.2. The finite-dimensional case

Let X be an orientable compact n-dimensional manifold without boundary, and let us consider an oriented vector bundle $\mathcal{E} \to X$ of rank $\text{rk}(\mathcal{E}) = 2m \leq n$ over X.

There are two complementary approaches to introducing the Euler class of $\mathcal{E}$, $e(\mathcal{E}) \in H^{2m}(X)$. The first one is in terms of a generic section, $s : X \to \mathcal{E}$. In that case $e(\mathcal{E})$ is constructed as the Poincaré dual in X of the homology class of $X_s = s^{-1}(0)$, the zero locus of s. When $\text{rk}(\mathcal{E}) = 2m = n = \dim(X)$, X_s is zero-dimensional (a set of points) and one can integrate $e(\mathcal{E})$ over X to obtain the Euler number of $\mathcal{E}$ as

$$\chi(\mathcal{E}) = \sum_{x_k : s(x_k)=0} (\pm 1), \tag{6.14}$$

i.e., one counts signs at the zeroes of a generic section s. When $2m < n$, X_s is generically $(n-2m)$-dimensional, and one can evaluate $e(\mathcal{E})$ on $2m$-cycles or equivalently take the product with elements of $H^{n-2m}(X)$ and evaluate it on X. That $e(\mathcal{E})$ is the Poincaré dual to X_s means that

$$\int_X e(\mathcal{E}) \wedge \mathcal{O} = \int_{X_s} i^* \mathcal{O}, \tag{6.15}$$

where $i : X_s \hookrightarrow X$ is the inclusion map, and $\mathcal{O}$ is a differential form representing a class in $H^{n-2m}(X)$. This result can be interpreted as a localization result, since it says that integrals on X involving $e(\mathcal{E})$ can be reduced to integrals over a submanifold X_s. Of course, as a cohomology class $e(\mathcal{E})$ does not depend on the choice of section, as long as it is generic. In the case in which the section is not generic the equality (6.15) is no longer true and has to be modified appropriately. When the bundle $\mathcal{E}$ is endowed with a connection ∇ the result can be stated as follows. Since $\nabla s \in \Omega^1(X, \mathcal{E})$ one can regard ∇s as a map:

$$\nabla s : T_p X \to \mathcal{E}_p, \tag{6.16}$$

where $p \in X$. If s is not generic, this map will have a non-trivial cokernel coker ∇s, and one can construct a bundle $\mathcal{E}'$ over X_s whose fiber at p is coker $(\nabla s)_p$. One then has:

$$\int_X e(\mathcal{E}) \wedge \mathcal{O} = \int_{X_s} e(\mathcal{E}') \wedge i^* \mathcal{O}. \tag{6.17}$$

This will be useful when discussing the extension of the Mathai–Quillen formalism to the infinite-dimensional case.

The second approach to the construction of the Euler class of $\mathcal{E}$ makes use of Chern–Weil theory, and gives a representative $e_\nabla(\mathcal{E})$ of $e(\mathcal{E})$ associated with a connection ∇ in $\mathcal{E}$:

$$e_\nabla(\mathcal{E}) = (2\pi)^{-m}\mathrm{Pf}(K), \tag{6.18}$$

where $\mathrm{Pf}(K)$ stands for the Pfaffian of the curvature K associated to the connection ∇. The representative $e_\nabla(\mathcal{E})$ can also be written as:

$$e_\nabla(\mathcal{E}) = (2\pi)^{-m}\int d\chi e^{\frac{1}{2}\chi_a K_{ab}\chi_b}, \tag{6.19}$$

by means of a set of real Grassmann-odd variables χ_a, $a = 1,\ldots,2m$, satisfying the Berezin rules of integration:

$$\int d\chi_a \chi_b = \delta_{ab}. \tag{6.20}$$

In this approach, when $\mathrm{rk}(\mathcal{E}) = 2m = n = \dim(X)$ the Euler number of $\mathcal{E}$ is computed by

$$\chi(\mathcal{E}) = \int_X e_\nabla(\mathcal{E}). \tag{6.21}$$

Again, $e_\nabla(\mathcal{E})$ is independent of the choice of connection ∇.

A third way to construct the Euler class of a vector bundle $\mathcal{E}$ is through its *Thom class*. The Thom class is characterized as follows. Let $\mathcal{E} \to X$ be a vector bundle of rank $2m$ with fiber $V = \mathbf{R}^{2m}$, and let us consider the cohomology of forms on $\mathcal{E}$ with compact support along the fiber. By integrating the form along the fiber one has an explicit isomorphism (the Thom isomorphism) between k forms over $\mathcal{E}$ and $k - 2m$ forms over X. This isomorphism can be made explicit with the aid of the Thom class, whose representative $\Phi(\mathcal{E})$ is a closed $2m$-form over $\mathcal{E}$ with compact support along the fiber such that its integral over the fiber is unity. If ω is an arbitrary p-form over X, its image under the Thom isomorphism is the $p + 2m$ form $\pi^*(\omega) \wedge \Phi(\mathcal{E})$, which by construction has compact support along the fiber. $\pi^*(\omega)$ is the pullback of ω by the projection $\pi : \mathcal{E} \to X$. If s is any section of $\mathcal{E}$, the pullback of the Thom form under s, $s^*\Phi(\mathcal{E})$, is a closed form in the same cohomology class as the Euler class $e(\mathcal{E})$. If s is a generic section then $s^*\Phi(\mathcal{E})$ is the Poincaré dual of the zero locus of s.

The Mathai–Quillen formalism provides an explicit representative of the Thom class $\Phi(\mathcal{E})$. The construction goes as follows: first, we assume that $\mathcal{E}$ is equipped with an inner product g compatible with the connection ∇, consequently

$$d(g(s,t)) = g(\nabla s, t) + g(s, \nabla t), \quad s,t \in \Gamma(\mathcal{E}). \tag{6.22}$$

As our vector bundle is oriented and has an inner product, we can reduce the structure group to $G = SO(2m)$. Let P be the principal G-bundle over X such that $\mathcal{E}$ is an associated vector bundle:

$$\mathcal{E} = P \times_G V. \tag{6.23}$$

In particular, $P \times V$ can be considered as a principal G-bundle over $\mathcal{E}$. Recall that when given a principal G-bundle $\pi : P \to X$, a differential form ϕ on P descends to a form on X if the following two conditions are satisfied: first, given vector fields V_i, $\phi(V_1, \dots, V_q) = 0$ whenever one of the V_i is vertical (and in this case ϕ is said to be *horizontal*). Second, ϕ is invariant under the G action. The forms that satisfy both conditions (*i.e.*, they are horizontal and invariant) are called *basic*. In particular, if we consider the principal bundle $P \times V \to P \times_G V$ we have an isomorphism:

$$\Omega^*(P \times_G V) \simeq \Omega^*(P \times V)_{\mathrm{basic}}. \tag{6.24}$$

Suppose now that P is endowed with a connection $\theta \in \Omega^1(P, \mathbf{g})$ and associated curvature K, and consider the Weil algebra of G, $\mathcal{W}(\mathbf{g})$. As $\mathbf{g} = so(2m)$ the generators are antisymmetric matrices θ_{ab} (of degree 1) and K_{ab} (of degree 2) (notice that we use the same notation for the connection and curvature of P and the generators of $\mathcal{W}(\mathbf{g})$). The property that $\mathcal{W}(\mathbf{g})$ provides a universal realization of the relations defining the curvature and connection on P gives the Chern–Weil homomorphism:

$$w : \mathcal{W}(\mathbf{g}) \longrightarrow \Omega^*(P), \tag{6.25}$$

defined in a natural way through the expansion of θ and K in (6.6) (for $G = SO(2m)$, the map w is just the correspondence between the generators of $\mathcal{W}(\mathbf{g})$ and the entries of the antisymmetric matrices for the curvature and connection in P). The Chern–Weil homomorphism maps the universal connection and curvature in the Weil algebra to the actual connection and curvature in P. Combined with the lifting of forms from V to $P \times V$, we obtain another homomorphism:

$$w \otimes \pi_2^* : \mathcal{W}(\mathbf{g}) \otimes \Omega^*(V) \longrightarrow \Omega^*(P \times V), \qquad (6.26)$$

where $\pi_2 : P \times V \to V$ is the projection on the second factor. It is easy to see that the basic subalgebra $\Omega_G(V)$ maps to the basic differential forms on $P \times V$, and therefore to the differential forms of the associated vector bundle $\mathcal{E}$. This is the geometric context of the Mathai–Quillen construction.

The universal Thom form $U(\mathcal{E})$ of Mathai and Quillen is an element in $\mathcal{W}(\mathbf{g}) \otimes \Omega^*(V)$ given by:

$$U(\mathcal{E}) = (2\pi)^{-m} \mathrm{Pf}\,(K) \exp\left(-\frac{1}{2}|v|^2 - \frac{1}{2}\nabla v_a (K^{-1})_{ab} \nabla v_b \right). \qquad (6.27)$$

In this expression the v_a are orthonormal coordinate functions on V, dv_a are their corresponding differentials, and $\nabla v_a = dv_a + \theta_{ab} v_b$. K and θ are the antisymmetric matrices of generators in $\mathcal{W}(\mathbf{g})$, and $|v|^2 = \sum_a v_a^2$. It is easy to write (6.27) as a Grassmann integral as in (6.19):

$$U(\mathcal{E}) = \int d\chi \, e^{-\frac{1}{2}|v|^2 + \frac{1}{2}\chi_a K_{ab}\chi_b + i\chi_a \nabla v_a}. \qquad (6.28)$$

One can check that $U(\mathcal{E})$ is a basic form, and so it belongs to $\Omega_G(V)$, and also that it is equivariantly closed. The image of $U(\mathcal{E})$ under the map (6.26)

$$\Phi(\mathcal{E}) = (w \otimes \pi_2^*)(U(\mathcal{E})) \qquad (6.29)$$

is then a basic closed differential form in $\Omega(P \times V)$ and it descends to a form in $\Omega^{2m}(\mathcal{E})$. Notice that $\Phi(\mathcal{E})$ has no compact support along the fibre, but rather a Gaussian decay. Nevertheless, one can define a cohomology of rapidly decreasing forms $H_{rd}^*(\mathcal{E})$ on the fibre V analogous to the cohomology of forms with compact support on V, and the usual results about the Thom class also hold in this slightly generalized setting. In particular, the Thom class can be uniquely characterized as a cohomology class in $H_{rd}^{2m}(\mathcal{E})$ such that its integration along the fibre is 1. This can be easily checked for the universal representative given in (6.27), therefore the image of $\Phi(\mathcal{E})$ in $H_{rd}^{2m}(\mathcal{E})$ is the Thom class of $\mathcal{E}$.

Let $s : X \to \mathcal{E}$ be a section of $\mathcal{E}$. Every section of $\mathcal{E}$ is associated with a G-equivariant map

$$\widehat{s} : P \to V, \qquad \widehat{s}(pg) = g^{-1}\widehat{s}(p). \qquad (6.30)$$

Consider now the map $\tilde{s} : P \to P \times V$ given by $\tilde{s}(p) = (p, \hat{s}(p))$. It follows that $\tilde{s}^*\Phi(P \times V)$ is a closed differential form on P which is also basic, therefore descends to a form on X which is precisely $e_{s,\nabla}(\mathcal{E})$. This form gives a representative of the Euler class of $\mathcal{E}$, and by using (6.28) we can also represent it as a Grassmann integral:

$$e_{s,\nabla}(\mathcal{E}) = (2\pi)^{-m} \int d\chi e^{-\frac{1}{2}|s|^2 + \frac{1}{2}\chi_a K_{ab}\chi_b + i\chi_a \nabla s_a}. \tag{6.31}$$

As a consistency check, note that, as follows from (6.19), $e_{s=0,\nabla}(\mathcal{E}) = e_\nabla(\mathcal{E})$, i.e., the pullback of the Mathai–Quillen form by the zero section gives back the Euler class of $\mathcal{E}$.

Let us denote by x^μ, $\mu = 1, \ldots, n$, a set of local coordinates on the base manifold X. The form $e_{s,\nabla}(\mathcal{E})$ can be rewritten in a compact way with the help of Grassmann odd real variables ψ^μ through the correspondence:

$$dx^\mu \leftrightarrow \psi^\mu,$$
$$\omega = \frac{1}{p!}\omega_{\mu_1,\ldots,\mu_p} dx^{\mu_1} \wedge \cdots \wedge dx^{\mu_p} \leftrightarrow \omega(\psi) = \frac{1}{p!}\omega_{\mu_1,\ldots,\mu_p}\psi^{\mu_1}\cdots\psi^{\mu_p}. \tag{6.32}$$

The integral over X of a top-form $\omega^{(n)}$ is therefore given by a simultaneous conventional integration over X and a Berezin integration over the ψs:

$$\int_X \omega^{(n)} = \int_X dx \int d\psi\, \omega^{(n)}(\psi). \tag{6.33}$$

In this language the Mathai–Quillen representative (6.31) can be rewritten as:

$$e_{s,\nabla}(\mathcal{E})(\psi) = (2\pi)^{-m} \int d\chi e^{-\frac{1}{2}|s|^2 + \frac{1}{2}\chi_a K_{ab}(\psi)\chi_b + i\chi_a \nabla s_a(\psi)}, \tag{6.34}$$

and, for example, in the case $n = 2m$ one has the following expression for the Euler number of $\mathcal{E}$:

$$\chi(\mathcal{E}) = (2\pi)^{-m} \int_X dx d\psi d\chi e^{-\frac{1}{2}|s|^2 + \frac{1}{2}\chi_a K_{ab}(\psi)\chi_b + i\chi_a \nabla s_a(\psi)}. \tag{6.35}$$

It is worthwhile remarking that (6.35) looks like the partition function of a field theory whose 'action' is:

$$S(x, \psi, \chi) = \frac{1}{2}|s|^2 - \frac{1}{2}\chi_a K_{ab}(\psi)\chi_b - i\chi_a \nabla s_a(\psi). \tag{6.36}$$

This action is invariant under the transformations:

$$\delta x^{\mu} = \psi^{\mu}, \qquad \delta \psi^{\mu} = 0, \qquad \delta \chi_a = i s_a. \qquad (6.37)$$

This suggests that the Mathai–Quillen formalism can be recast in a more conventional physical language through the use of the BRST formalism, which was created originally to quantize Yang–Mills theories. To do that we first introduce an auxiliary bosonic field B_a, which has the meaning of a basis of differential forms for the fiber. In this way we obtain two pairs of fields (x, ψ) and (χ, B), where the first pair is associated with the base manifold X, whilst the second pair is associated to the fiber V. Notice that these pairs have opposite Grassmannian character. We also extend the transformation δ in (6.37) to the new set of fields as follows:

$$\begin{aligned} \delta x^{\mu} &= \psi^{\mu}, & \delta \psi^{\mu} &= 0, \\ \delta \chi_a &= B_a, & \delta B_a &= 0. \end{aligned} \qquad (6.38)$$

Notice that $\delta^2 = 0$. Indeed, δ is simply the de Rham differential for the base manifold and for the fiber. One can then verify, by integrating out the auxiliary field B, that the Euler class can be written as the integral

$$e_{s,\nabla}(\mathcal{E}) = \frac{1}{(2\pi)^{2m}} \int dx dB e^{-\delta \Psi_{\text{loc}}}, \qquad (6.39)$$

where the so called localizing gauge fermion Ψ_{loc} is given by

$$\Psi_{\text{loc}} = \chi_a \left(i s_a + \frac{1}{2} \theta_{ab}(\psi) \chi_b + \frac{1}{2} B_a \right). \qquad (6.40)$$

The Mathai–Quillen representative interpolates between the two different approaches to the Euler class of a vector bundle which we explained at the beginning of this section. This statement can be made more precise as follows. The construction of $e_{s,\nabla}(\mathcal{E})$ is such that it is cohomologous to $e_{\nabla}(\mathcal{E})$ for any choice of a generic section s. Take, for example, the case $n = 2m$ and rescale $s \to \gamma s$. Nothing should change, so in particular:

$$\chi(\mathcal{E}) = \int_X e_{\gamma s, \nabla}(\mathcal{E}). \qquad (6.41)$$

We can now study (6.41) in two different limits. In the first limit, $\gamma \to 0$, after using (6.19) $\chi(\mathcal{E}) = (2\pi)^{-m} \int \text{Pf}(\Omega_{\nabla})$. In the second limit, $\gamma \to \infty$, the curvature term in (6.35) can be neglected, leading to $\chi(\mathcal{E}) = \sum_{x_k : s(x_k) = 0} (\pm 1)$.

These signs are generated by the ratio of the determinants of ∇s and its modulus, which results from the Gaussian integrations after expanding around each zero x_k. Hence we recover from this unified point of view the two complementary ways of defining the Euler class described at the beginning of the section.

We should point out that from the point of view of the BRST formalism which we introduced above, the independence of the integral with respect to γ is a consequence of the δ-exactness of the action (6.36). In particular, that the integral can be evaluated at $\gamma \to \infty$ means that the semi-classical approximation is exact, and that the integral localizes to the minima of the action $S(x, \psi, \chi)$, which in this case are the zeroes of the section, $s = 0$. This is then a finite-dimensional version of the functional integral computation that we did in the previous chapter, in the context of Donaldson–Witten theory.

6.3. A detailed example

To further clarify the formalism let us work out an explicit example. To be definite we will consider $\mathcal{E} = TX$. The section s is taken as a vector field V on X, which we assume to be generic. After rescaling $V \to \gamma V$ the action (6.36) takes the form:

$$S(x, \psi, \chi) = \frac{1}{2}\gamma^2 g_{\mu\nu} V^\mu V^\nu - \frac{1}{4}\chi_a R^{ab}{}_{\mu\nu}\psi^\mu\psi^\nu\chi_b - i\gamma\nabla_\mu V^\nu\psi^\mu e^a_\nu\chi_a. \quad (6.42)$$

As we explained in the previous section, it is possible to introduce auxiliary fields in the formulation. In the example under consideration we use

$$e^{-\frac{1}{2}\gamma^2 g_{\mu\nu} V^\mu V^\nu} = \frac{\gamma^{2m}\sqrt{g}}{(2\pi)^m}\int dB e^{-\frac{1}{2}\gamma^2(g_{\mu\nu} B^\mu B^\nu + 2iB_\nu V^\nu)}, \quad (6.43)$$

where B^μ is an auxiliary field. The Euler number resulting from (6.42) can be rewritten as:

$$\chi(X) = \int_X dx d\psi d\chi dB$$

$$\frac{\gamma^{2m}\sqrt{g}}{(2\pi)^{2m}}e^{-\frac{1}{2}\gamma^2(g_{\mu\nu}B^\mu B^\nu + 2iB_\nu V^\nu) + \frac{1}{4}\chi_a R^{ab}{}_{\mu\nu}\psi^\mu\psi^\nu\chi_b + i\gamma\nabla_\mu V^\nu\psi^\mu e^a_\nu\chi_a}.$$

$$(6.44)$$

Making the redefinitions:

$$\psi^\mu \to \gamma^{\frac{1}{2}}\psi^\mu, \qquad \chi_a \to \gamma^{\frac{1}{2}}e_{a\mu}\overline{\psi}^\mu,$$

$$d\psi \to (\gamma^{-\frac{1}{2}})^{2m}d\psi, \qquad d\chi \to (\gamma^{-\frac{1}{2}})^{2m}\frac{1}{\sqrt{g}}d\overline{\psi}, \qquad (6.45)$$

we obtain:

$$\chi(X) = \int_X dx d\psi d\overline{\psi} dB$$

$$\frac{1}{(2\pi)^{2m}} e^{-\gamma^2 \left(\frac{1}{2} g_{\mu\nu} (B^\mu B^\nu + 2i B^\mu V^\nu) - \frac{1}{4} R^{\rho\sigma}{}_{\mu\nu} \overline{\psi}_\rho \overline{\psi}_\sigma \psi^\mu \psi^\nu - i \nabla_\mu V^\nu \psi^\mu \overline{\psi}_\nu \right)}.$$

$$(6.46)$$

As we pointed out above, this looks like the partition function of a topological quantum field theory in which $g = 1/\gamma$ plays the role of the coupling constant. Furthermore, the exponent of (6.46) is invariant under the symmetry:

$$\delta x^\mu = \psi^\mu, \qquad\qquad \delta\overline{\psi}_\mu = B_\mu,$$
$$\delta\psi^\mu = 0, \qquad\qquad \delta B_\mu = 0. \qquad\qquad (6.47)$$

This symmetry agrees with the general structure presented in (6.38), and $(\overline{\psi}_\mu, B_\mu)$ is the doublet associated with the fiber, which in this case has the same dimension as the base manifold X. One easily finds that the exponent of (6.47) is indeed δ-exact:

$$\chi = \int_X dx d\psi d\overline{\psi} dB \frac{1}{(2\pi)^{2m}} e^{-\gamma^2 \delta \left(\frac{1}{2} \overline{\psi}_\mu (B^\mu + 2i V^\mu + \Gamma^\sigma_{\tau\nu} \overline{\psi}_\sigma \psi^\nu g^{\mu\tau}) \right)}. \qquad (6.48)$$

Notice that the action in (6.48) has the general structure $\delta\Psi_{\mathrm{loc}}$, where the localizing gauge fermion is indeed of the form (6.40).

The fact that the action is δ-exact makes it possible to use field-theoretic arguments similar to those explained in the previous chapter to conclude that χ is independent of the coupling γ and of the metric $g_{\mu\nu}$. We can for example compute (6.35) in the semi-classical limit $\gamma \to \infty$, which in this case is exact. First, we expand around the zeroes, x_k, of V ($V^\mu(x_k) = 0$):

$$\chi(X) = \sum_{x_k} \int_X dx d\psi d\chi$$

$$(2\pi)^{-m} e^{-\frac{1}{2}\gamma^2 g_{\mu\nu} \partial_\sigma V^\mu \partial_\rho V^\nu x^\sigma x^\rho + \frac{1}{4} \chi_a R^{ab}{}_{\mu\nu} \psi^\mu \psi^\nu \chi_b + i\gamma \partial_\mu V^\nu \psi^\mu e^a_\nu \chi_a}.$$

$$(6.49)$$

Next, we rescale the variables in the following way:

$$x \to \gamma^{-1} x, \qquad\qquad d\psi \to \gamma^{\frac{1}{2}} d\psi,$$
$$dx \to \gamma^{-1} dx, \qquad\qquad \chi \to \gamma^{-\frac{1}{2}} \chi, \qquad\qquad (6.50)$$
$$\psi \to \gamma^{-\frac{1}{2}} \psi, \qquad\qquad d\chi \to \gamma^{\frac{1}{2}} d\chi.$$

The measure is invariant under this rescaling. Using the shorthand notation for the Hessian $H_\sigma^{(k)\mu} = \partial_\sigma V^\mu\big|_{x_k}$, one finds after taking the limit $\gamma \to \infty$:

$$\chi(X) = \sum_{x_k} \int_X dx d\psi d\chi (2\pi)^{-m} e^{-\frac{1}{2}g_{\mu\nu}H_\sigma^{(k)\mu}H_\rho^{(k)\nu}x^\sigma x^\rho + iH_\mu^{(k)\nu}\psi^\mu e_\nu^a \chi_a}$$

$$= \sum_{x_k} (2\pi)^{-m} \frac{(\sqrt{2\pi})^{2m}}{\sqrt{g}|\det H^{(k)}|} \frac{e}{(2m)!} (2m)! \det H^{(k)} = \sum_{x_k} \frac{\det H^{(k)}}{|\det H^{(k)}|},$$

$$(6.51)$$

which indeed corresponds to the Euler number of X by virtue of the Poincaré–Hopf theorem. Notice that the computation in the semi-classical limit gives a quotient of determinants which is ± 1. This is because there is a hidden supersymmetry in the action, as it is manifest in the BRST formulation, and we have a cancellation of fermionic and bosonic determinants. The same cancellation appeared in the analysis of the functional integral in Donaldson–Witten theory.

Let us further illustrate the example (6.42). Consider the two-sphere $\mathbf{S}^2$ with the standard parametrization:

$$\alpha : (0, \pi) \times (0, 2\pi) \longrightarrow \mathbf{R}^3,$$
$$(\theta, \varphi) \longrightarrow (\sin\theta\cos\varphi, \sin\theta\sin\varphi, \cos\theta). \tag{6.52}$$

In terms of these coordinates we have the relations:

$$ds^2 = d\theta^2 + \sin^2\theta d\varphi^2, \qquad g_{\mu\nu} = \begin{pmatrix} 1 & 0 \\ 0 & \sin^2\theta \end{pmatrix}, \tag{6.53}$$

and the following values for the Christoffel symbol ($\Gamma_{\mu\nu}^\lambda = \frac{1}{2}g^{\lambda\sigma}(\partial_{(\mu}g_{\nu)\sigma} - \partial_\sigma g_{\mu\nu})$),

$$\Gamma_{\theta\theta}^\theta = \Gamma_{\varphi\varphi}^\varphi = \Gamma_{\theta\theta}^\varphi = \Gamma_{\theta\varphi}^\theta = 0,$$
$$\Gamma_{\varphi\varphi}^\theta = -\sin\theta\cos\theta, \qquad \Gamma_{\theta\varphi}^\varphi = \frac{\cos\theta}{\sin\theta}. \tag{6.54}$$

Let us pick an orthonormal frame:

$$e_a^\mu = \begin{pmatrix} 1 & 0 \\ 0 & \frac{1}{\sin\theta} \end{pmatrix}, \qquad e_\mu^a = \begin{pmatrix} 1 & 0 \\ 0 & \sin\theta \end{pmatrix}, \tag{6.55}$$

where the vierbeins satisfy the standard relations: $e_a^\mu e_b^\nu g_{\mu\nu} = \delta_{ab}$, $e_\mu^a e_\nu^b \delta_{ab} = g_{\mu\nu}$. The Riemann curvature tensor ($R^\lambda{}_{\mu\nu\kappa} = \partial_{[\kappa}\Gamma_{\nu]\mu}^\lambda + \Gamma_{\mu[\nu}^\tau \Gamma_{\kappa]\tau}^\lambda$) in (θ, φ) coordinates is given by:

$$R^\theta{}_{\varphi\varphi\theta} = \sin^2\theta, \tag{6.56}$$

while the curvature two-form Ω^{ab} takes the form:

$$\Omega^{12} = R^{12}{}_{\varphi\theta}d\varphi \wedge d\theta = e^1_\theta e^2_\varphi g^{\varphi\varphi} R^\theta{}_{\varphi\varphi\theta}d\varphi \wedge d\theta = \sin\theta d\varphi \wedge d\theta. \quad (6.57)$$

Next let us consider the vector field:

$$V^a = (\sin\varphi, \cos\varphi\cos\theta) \rightarrow V^\mu = (\sin\varphi, \cos\varphi\cot\theta). \quad (6.58)$$

This vector field has zeros at $\varphi = 0$, $\theta = \pi/2$ and $\varphi = \pi$, $\theta = \pi/2$. The components of the form ∇V^a are:

$$\nabla_\theta V^\varphi = -\cos\varphi, \quad \nabla_\varphi V^\theta = \sin^2\theta\cos\varphi, \quad \nabla_\theta V^\theta = \nabla_\varphi V^\varphi = 0, \quad (6.59)$$

or, alternatively,

$$\nabla_\theta V^a = e^a_\mu \nabla_\theta V^\mu = (0, -\cos\varphi\sin\theta), \qquad \nabla_\varphi V^a = e^a_\mu \nabla_\varphi V^\mu = (\cos\varphi\sin\theta, 0), \quad (6.60)$$

and therefore

$$\nabla V^a = (\sin^2\theta\cos\varphi d\varphi, -\cos\varphi\sin\theta d\theta). \quad (6.61)$$

The Euler class representative

$$e_{V,\nabla}(T\mathbf{S}^2) = \frac{1}{2\pi}\int d\chi_1 d\chi_2 e^{-\frac{1}{2}V^a V^a + \frac{1}{2}\chi_a \Omega^{ab}_\nabla \chi_b + i\nabla V^a \chi_a}, \quad (6.62)$$

after performing the rescaling $V^a \rightarrow \gamma V^a$,

$$-\frac{1}{2}V^a V^a + \frac{1}{2}\chi_a \Omega^{ab}_\nabla \chi_b + i\nabla V^a \chi_a$$
$$\longrightarrow -\frac{\gamma^2}{2}(\sin^2\theta + \cos^2\varphi\cos^2\theta) - \chi_1\chi_2 \sin\theta d\theta \wedge d\varphi \quad (6.63)$$
$$+ i\gamma(\sin^2\theta\cos\varphi d\varphi\chi_1 - \cos\varphi\sin\theta d\theta\chi_2),$$

becomes:

$$e_{V,\nabla}(T\mathbf{S}^2) = \frac{1}{2\pi}e^{-\frac{\gamma^2}{2}(\sin^2\theta + \cos^2\varphi\cos^2\theta)}\sin\theta(1 + \gamma^2\cos^2\varphi\sin^2\theta)d\theta \wedge d\varphi. \quad (6.64)$$

The Euler number of $\mathbf{S}^2$ is given by the integral:

$$\chi(\mathbf{S}^2) = \frac{1}{2\pi}\int_0^{2\pi} d\varphi \int_0^\pi d\theta \sin\theta e^{-\frac{\gamma^2}{2}(\sin^2\theta + \cos^2\varphi\cos^2\theta)}(1 + \gamma^2\cos^2\varphi\sin^2\theta). \quad (6.65)$$

Although γ appears explicitly in this expression, the result of the integration should be independent of γ. The reader is urged to prove it. Here we will perform two independent checks. On the one hand, in the limit $\gamma \to 0$ (6.65) gives trivially the correct result, $\chi(\mathbf{S}^2) = 2$. On the other hand, one can explore the opposite limit $\gamma \to \infty$, where the integral

$$\chi(\mathbf{S}^2) = \int_{S^2} dx \int d\psi^1 d\psi^2 \int d\chi_1 d\chi_2 \frac{1}{2\pi} e^{-A(x,\psi,\chi)}, \qquad (6.66)$$

with $S(x, \psi, \chi) = \frac{1}{2} V^a V^a - \frac{1}{2} \chi_a \Omega^{ab}_{\nabla} \chi_b - i \nabla V^a \chi_a$, is dominated by the zeros of V^a: $(\theta, \phi) = (\pi/2, 0)$ and $(\theta, \phi) = (\pi/2, \pi)$. This is the semi-classical approximation to the evaluation of the integral, which in this case is exact. The zeroes of the vector field are minima of the action, and we expand about them:

(a) $\theta = \pi/2 + x$, $\varphi = 0 + y$.

$$V^a V^a = \sin^2 \varphi + \cos^2 \varphi \cos^2 \theta = x^2 + y^2 + \cdots$$

$$\frac{1}{4} \chi_a R^{ab}_{\ \ \mu\nu} \psi^\mu \psi^\nu \chi_b = \chi_1 \chi_2 \psi^\theta \psi^\varphi R^{12}_{\ \ \theta\varphi} = \sin\theta \chi_1 \chi_2 \psi^\theta \psi^\varphi$$

$$= \left(1 - \frac{x^2}{2} + \cdots\right) \chi_1 \chi_2 \psi^\theta \psi^\varphi \qquad (6.67)$$

$$\nabla_\mu V^\nu \psi^\mu e^a_\nu \chi_a = -\cos\varphi \sin\theta \psi^\theta \chi_2 + \cos\varphi \sin^2\theta \psi^\varphi \chi_1$$

$$= -\psi^\theta \chi_2 + \psi^\varphi \chi_1 + \cdots,$$

and after performing the rescaling

$$x, y \to \gamma^{-1} x, \gamma^{-1} y,$$

$$\psi^\mu \to \gamma^{-\frac{1}{2}} \psi^\mu, \qquad (6.68)$$

$$\chi_a \to \gamma^{-\frac{1}{2}} \chi_a,$$

one obtains:

$$\frac{1}{2\pi} \int dx dy \, e^{-\frac{1}{2}(x^2+y^2)} \int d\chi_1 d\chi_2 d\psi^\theta d\psi^\varphi e^{i(-\psi^\theta \chi_2 + \psi^\varphi \chi_1)} = 1. \qquad (6.69)$$

(b) $\theta = \pi/2 + x$, $\varphi = \pi + y$.

$$\frac{1}{2\pi} \int dx dy \, e^{-\frac{1}{2}(x^2+y^2)} \int d\chi_1 d\chi_2 d\psi^\theta d\psi^\varphi e^{i(\psi^\theta \chi_2 - \psi^\varphi \chi_1)} = 1. \qquad (6.70)$$

Notice that in the limit $\gamma \to \infty$ we have reproduced the pattern described in (6.51):

$$\chi(\mathbf{S}^2) = \left(\frac{\det \begin{pmatrix} 0 & -1 \\ 1 & 0 \end{pmatrix}}{\left| \det \begin{pmatrix} 0 & -1 \\ 1 & 0 \end{pmatrix} \right|} + \frac{\det \begin{pmatrix} 0 & 1 \\ -1 & 0 \end{pmatrix}}{\left| \det \begin{pmatrix} 0 & 1 \\ -1 & 0 \end{pmatrix} \right|} \right) = 2. \qquad (6.71)$$

6.4. Mathai–Quillen formalism: Infinite-dimensional case

In this section we consider the extension of the Mathai–Quillen formalism to the infinite-dimensional case, *i.e.*, to infinite-dimensional vector bundles over infinite-dimensional configuration spaces. Of course, in these cases $e(\mathcal{E})$ is not defined as a geometric object. However, one can do the following: let us consider a section $s : \mathcal{X}/\mathcal{G} \to \mathcal{E}$ which has a *finite-dimensional* zero locus $\mathcal{M}$, and let us use the Mathai–Quillen formalism to give a formal construction of $e_{s,\nabla}(\mathcal{E})$. We can now take into account (6.15) to *define* integrals over the configuration space $\mathcal{X}/\mathcal{G}$ as follows:

$$\int_{\mathcal{X}/\mathcal{G}} e_{s,\nabla}(\mathcal{E}) \wedge \mathcal{O} = \int_{\mathcal{M}} i^* \mathcal{O}, \qquad (6.72)$$

where $\mathcal{O} \in H^d(\mathcal{X}/\mathcal{G})$ and $d = \dim \mathcal{M}$. If the section s is not generic one has to use (6.17) instead of (6.15) in order to define the left hand side of (6.72), and we have

$$\int_{\mathcal{X}/\mathcal{G}} e_{s,\nabla}(\mathcal{E}) \wedge \mathcal{O} = \int_{\mathcal{M}} e(\mathcal{E}') \wedge i^* \mathcal{O}, \qquad (6.73)$$

where $\mathcal{E}'$ is the bundle over $\mathcal{M}$ whose fiber is coker ∇s. Since we are defining the integrals over the configuration space by using (6.72) and (6.73), they depend *a priori* on the particular section chosen. However, one expects that this is still a way of obtaining interesting topological invariants, and the known examples —such us Donaldson–Witten theory— show that this is indeed the case. As we will see, in the infinite-dimensional case the left hand side of (6.72) or (6.73) becomes a correlation function of a topological quantum field theory, and the definition of these correlation functions as finite-dimensional integrals agrees with the quantum field theory analysis. For example, we saw in the previous chapter that the functional integral of Donaldson–Witten theory reduces to a finite-dimensional integral over zero-modes. The Mathai–Quillen formalism provides a geometric explanation of why this should be the

case, and shows that for TQFTs of cohomological type, functional integrals
can be given a precise meaning in terms of finite-dimensional integrals over a
moduli space.

Another observation which will be useful is the following: the tangent
space of $\mathcal{M}$ is given by $\ker \nabla s$, and $\dim(\mathcal{M}) = \dim \ker \nabla s$. Therefore the
integral in the right hand side of (6.73) will only be different from zero if
the degree of $\mathcal{O}$ is equal to $\operatorname{ind} \nabla s = \dim \ker \nabla s - \dim \operatorname{coker} \nabla s$. Of course
$\dim(\mathcal{M}) = \operatorname{ind} \nabla s$ for a generic section. The index $\operatorname{ind} \nabla s$ is usually called
the virtual dimension of the moduli space. Notice that this is exactly the
situation we found in the analysis of Donaldson–Witten theory in Chapters 2
and 5. The virtual dimension coincides with the actual dimension only when
$\dim \operatorname{coker} \nabla s = 0$ and the section is generic. When $\operatorname{ind} \nabla s = 0$ we can use
(6.72) or (6.73) with $\mathcal{O} = 1$ to define the *regularized* Euler number of $\mathcal{E}$, $\chi_s(\mathcal{E})$.
This corresponds in field-theoretic language to the partition function of the
model.

The requirement on the degree of $\mathcal{O}$ can be made more transparent if
we introduce a ghost number assignment to the different fields involved in
$e_{s,\nabla}(\mathcal{E})$, as follows:

$$x \longrightarrow 0, \qquad \psi \longrightarrow 1, \qquad \chi \longrightarrow -1, \qquad B \longrightarrow 0. \qquad (6.74)$$

Notice that the transformation δ has ghost number 1, and the action $S = \delta \Psi_{\text{loc}}$ has ghost number zero. Interpreting δ as a $U(1)$ symmetry of S, we see
that $\operatorname{ind} \nabla s$ is the anomaly of that symmetry, and the requirement that the
degree of $\mathcal{O}$ is equal to $\operatorname{ind} \nabla s$ is just the requirement that the insertion of
operators in the functional integral (6.72) soaks up the zero-modes associated
to the anomaly in order to obtain a non-vanishing correlation function. This
is exactly what we found in our analysis of Donaldson–Witten theory, but it is
a general feature of cohomological field theories and, as we have just seen, can
be understood in a clear geometric way in the context of the Mathai–Quillen
formalism.

The construction of $e_{s,\nabla}(\mathcal{E})$ follows the pattern of the finite-dimensional
case and it will be illustrated by the description of two examples. The first
one corresponds to supersymmetric quantum mechanics. This is the simplest
infinite-dimensional case and leads to an interesting reformulation of singularity theory. The second example deals with topological sigma models. None
of these examples involve gauge symmetries. In other words, the group of

symmetries of the configuration space $\mathcal{G}$ will be taken to be trivial. In the case that the moduli problem has a gauge symmetry we shall need a slight reformulation of the Mathai–Quillen formalism which will be discussed in the next section.

1) Supersymmetric quantum mechanics

Let X be a smooth, orientable, Riemannian manifold with metric $g_{\mu\nu}$. The configuration space of the theory $\mathcal{X}$ will be the loop space LX, which is defined by the set of smooth maps:

$$x : \mathbf{S}^1 \to X,$$
$$t \in [0,1] \to x^\mu(t), \qquad (6.75)$$
$$x^\mu(0) = x^\mu(1).$$

The vector bundle $\mathcal{E}$ will be in this case $T(LX)$, the tangent vector bundle of LX, with fiber $\mathcal{F} = T_x(LX) = \Gamma(x^*(TX))$. A section of $\mathcal{E}$ is a vector field over LX, and it has the form:

$$V(x) = \oint dt V^\mu(x(t)) \frac{\partial}{\partial x^\mu(t)}. \qquad (6.76)$$

The metric on X provides a natural metric for $T_x(LX)$: let V_1, $V_2 \in T_x(LX)$, then

$$\widehat{g}_x(V_1, V_2) = \oint dt g_{\mu\nu}(x(t)) V_1^\mu(x(t)) V_2^\nu(x(t)). \qquad (6.77)$$

The Levi–Civita connection on LX is the pullback connection from X:

$$\nabla V = \oint dt_1 dt_2 \left[\frac{\delta V^\mu(x(t_1))}{\delta x^\nu(t_2)} + \Gamma^\mu_{\nu\lambda}(x(t_2)) V^\lambda(x(t_1)) \delta(t_1 - t_2) \right]$$
$$\times \frac{\partial}{\partial x^\mu(t_1)} \otimes \widehat{dx}^\nu(t_2), \qquad (6.78)$$

where $\{\partial/\partial x^\mu(t)\}$ is a basis of $T_x(LX)$, and $\{\widehat{dx}^\nu(t)\}$ is a basis of $T_x^*(LX)$. Let us consider the vector field:

$$V^\mu(x) = \frac{d}{dt} x^\mu \equiv \dot{x}^\mu. \qquad (6.79)$$

The Mathai–Quillen representative of $e_{s,\nabla}(\mathcal{E})$ is constructed following the same procedure as in the finite-dimensional case:

Finite–
dimensional
case

$$
\begin{cases}
\chi = \displaystyle\int_X dx\, d\psi\, d\overline{\psi}\, dB \,\frac{1}{(2\pi)^{2m}} \mathrm{e}^{-\gamma^2 \delta\Psi(x,\psi,\overline{\psi},B)}, \\[2mm]
\Psi(x,\psi,\overline{\psi},B) = \dfrac{1}{2}\overline{\psi}_\mu(B^\mu + 2iV^\mu + \Gamma^\sigma_{\tau\nu}\overline{\psi}_\sigma \psi^\nu g^{\mu\tau}), \\[2mm]
\delta x^\mu = \psi^\mu, \qquad \delta\psi^\mu = 0, \\[2mm]
\delta\overline{\psi}_\mu = B_\mu, \qquad \delta B_\mu = 0.
\end{cases}
$$

Supersymmetric
quantum
mechanics

$$
\begin{cases}
\chi_V = \displaystyle\int_{LX} dx\, d\psi\, d\overline{\psi}\, dB \,\frac{1}{(2\pi)^{2m}} \mathrm{e}^{-\gamma^2 \delta\Psi(x,\psi,\overline{\psi},B)}, \\[2mm]
\Psi(x,\psi,\overline{\psi},B) = \dfrac{1}{2}\oint dt\, \overline{\psi}_\mu(B^\mu + 2i\dot{x}^\mu + \Gamma^\sigma_{\tau\nu}\overline{\psi}_\sigma \psi^\nu g^{\mu\tau}), \\[2mm]
\delta x^\mu(t) = \psi^\mu(t), \qquad \delta\psi^\mu(t) = 0, \\[2mm]
\delta\overline{\psi}_\mu(t) = B_\mu(t), \qquad \delta B_\mu(t) = 0.
\end{cases}
$$

$$(6.80)$$

We can now compute the regularized Euler number of $\mathcal{E}$, $\chi_V(\mathcal{E})$, following the observations above. First of all, by using the Mathai–Quillen representative $\chi_V(\mathcal{E})$ can be written as a functional integral. After we integrate out the auxiliary fields in the action in (6.80), this integral reads:

$$
\chi - V(\mathcal{E}) = \int \frac{dx\, d\psi\, d\overline{\psi}}{(2\pi)^m} \frac{1}{\sqrt{g}\gamma^{2m}} \mathrm{e}^{-\gamma^2 \oint dt\left[\frac{1}{2}g_{\mu\nu}\dot{x}^\mu \dot{x}^\nu - i\overline{\psi}\nabla_t \psi - \frac{1}{4}R^{\rho\sigma}{}_{\mu\nu}\overline{\psi}_\rho \overline{\psi}_\sigma \psi^\mu \psi^\nu\right]},
$$

$$(6.81)$$

where ∇t is the operator

$$
\nabla_t \psi^\mu = \dot{\psi}^\mu + \Gamma^\mu_{\nu\sigma}\psi^\sigma \dot{x}^\nu.
$$

$$(6.82)$$

The action in the exponential of (6.81) is precisely the action corresponding to supersymmetric quantum mechanics. The δ-transformations become

$$
\delta x^\mu = \psi^\mu, \qquad \delta\psi^\mu = 0, \qquad \delta\overline{\psi}_\mu = -ig_{\mu\nu}\dot{x}^\nu - \Gamma^\sigma_{\mu\nu}\overline{\psi}_\sigma \psi^\nu,
$$

$$(6.83)$$

which close only on-shell, i.e., $\delta^2 = 0$ modulo field equations (of course, by reintroducing the vector field B one can close δ off-shell). As discussed below, these transformations can be regarded as supersymmetry transformations. Following (6.74) we introduce the ghost number symmetry

$$
x \longrightarrow 0, \qquad \psi \longrightarrow 1, \qquad \overline{\psi} \longrightarrow -1, \qquad B \longrightarrow 0.
$$

$$(6.84)$$

One can show that $\operatorname{ind} \nabla_t = 0$.

Let us now evaluate $\chi_V(\mathcal{E})$. Geometrically the Euler number of $\mathcal{E}$ is not well defined, since we are in an infinite-dimensional setting. However, we can use (6.72) or (6.73) to evaluate the regularized Euler number $\chi_V(\mathcal{E})$, depending on whether the section is generic or not. The zero locus of the vector field V in (6.79) is the space of constant loops $\mathcal{M} = X$. However, this section is not generic, since over $p \in X$ we have $\ker \nabla_t = \operatorname{coker} \nabla_t = T_p X$. We then have by using (6.73)

$$\chi_V(\mathcal{E}) = \chi(TX), \tag{6.85}$$

and the regularized Euler characteristic of $\mathcal{E}$ is in this case the Euler characteristic of X.

We can check that the definition of $\chi_V(\mathcal{E})$ provided by (6.73) agrees with our physical expectations by computing the functional integral (6.81) in the limit $\gamma \to \infty$. In this limit the exact result is obtained very simply by considering the expansion of the exponential around bosonic and fermionic zero-modes:

Bosonic part: $\dot{x}^\mu = 0 \to x^\mu$ constant

Fermionic part: $\begin{cases} \psi^\mu(t) = \psi^\mu + \text{non-zero-modes} \\ \overline{\psi}^\mu(t) = \overline{\psi}^\mu + \text{non-zero-modes} \end{cases}$

The integration over the non-zero-modes is trivial since the δ symmetry implies that the ratio of determinants is equal to 1. The integration over the zero-modes gives:

$$\begin{aligned}
\chi_V &= \int_X dx \frac{(2\pi)^{-m}}{\gamma^{2m}\sqrt{g}} \int \left[\prod_{\mu=1}^{2m} d\psi^\mu \right] \left[\prod_{\nu=1}^{2m} d\overline{\psi}^\nu \right] e^{\gamma^2 \frac{1}{4} R^{\rho\sigma}{}_{\mu\nu} \overline{\psi}_\rho \overline{\psi}_\sigma \psi^\mu \psi^\nu} \\
&= \int_X \frac{1}{\gamma^{2m}(2\pi)^m} \int \left(\prod_{a=1}^{2m} d\chi_a \right) e^{\frac{\gamma^2}{2} \chi_a \Omega^{ab} \chi_b} = \int_X \frac{1}{(2\pi)^m} \operatorname{Pf}(\Omega^{ab}) \\
&= \chi(X),
\end{aligned} \tag{6.86}$$

where $\chi_a = e^\mu_a \overline{\psi}_\mu$.

To finish this quick tour through supersymmetric quantum mechanics, it is interesting to recall that χ_V can be also computed using Hamiltonian methods. The expression (6.81) possesses a second δ-like symmetry, $\bar{\delta}$:

$$\bar{\delta}x^{\mu} = \bar{\psi}^{\mu},$$
$$\bar{\delta}\bar{\psi}_{\mu} = 0, \qquad\qquad\qquad (6.87)$$
$$\bar{\delta}\psi_{\mu} = -ig_{\mu\nu}\dot{x}^{\nu} - \Gamma^{\sigma}_{\mu\nu}\psi_{\sigma}\bar{\psi}^{\nu}.$$

After using the field equations one finds that

$$\delta^2 = 0, \qquad \bar{\delta}^2 = 0, \qquad \delta\bar{\delta} + \bar{\delta}\delta = \frac{d}{dt}, \qquad (6.88)$$

which, in terms of operators,

$$\delta \leftrightarrow Q, \qquad \bar{\delta} \leftrightarrow \bar{Q}, \qquad \frac{d}{dt} \leftrightarrow H, \qquad (6.89)$$

(H stands for the Hamiltonian operator) implies that:

$$Q^2 = \bar{Q}^2 = 0, \qquad \{Q, \bar{Q}\} = Q\bar{Q} + \bar{Q}Q = H, \qquad (6.90)$$

which is the standard supersymmetry algebra for $(0+1)$-supersymmetric field theories. We can carry out explicitly the canonical quantization of the theory by imposing the canonical commutation relations:

$$\{\bar{\psi}^{\mu}, \psi^{\nu}\} = g^{\mu\nu}, \qquad \{\psi^{\mu}, \psi^{\nu}\} = \{\bar{\psi}^{\mu}, \bar{\psi}^{\nu}\} = 0. \qquad (6.91)$$

From these equations it is natural to interpret $\bar{\psi}$ as fermion creation operators. In view of this we have the following structure on the Hilbert space:

$$\left\{ \begin{array}{l} \text{States with one fermion: } \omega_{\mu}(x)\bar{\psi}^{\mu}|\Omega\rangle, \\[2mm] \text{States with two fermions: } \omega_{\mu\nu}(x)\bar{\psi}^{\mu}\bar{\psi}^{\nu}|\Omega\rangle, \\[1mm] \quad\vdots \\[1mm] \text{States with } n \text{ fermions: } \omega_{\mu_1,\ldots,\mu_n}(x)\bar{\psi}^{\mu_1}\cdots\bar{\psi}^{\mu_n}|\Omega\rangle, \end{array} \right. \qquad (6.92)$$

$|\Omega\rangle$ being the Clifford vacuum. The Hilbert space of our system is thus $\Omega^*(X)$, the set of differential forms on X. On this Hilbert space Q and $\bar{Q}$ are represented by the exterior derivative d and its adjoint d^+

$$Q \leftrightarrow d, \qquad \bar{Q} \leftrightarrow d^+, \qquad (6.93)$$

therefore, the Hamiltonian is the Hodge–de Rham Laplacian on X:

$$H = dd^+ + d^+d = \Delta. \qquad (6.94)$$

The zero energy states are in one-to-one correspondence with the harmonic forms on X. After rescaling the parameter t and the fermionic fields by

$$t \to \gamma^2 t, \qquad \overline{\psi} \to \gamma^{-1}\overline{\psi}, \qquad \psi \to \gamma^{-1}\psi, \qquad (6.95)$$

the partition function (6.81) takes the form:

$$\chi_V(\mathcal{E}) = \int \frac{dx\,d\psi\,d\overline{\psi}}{(2\pi)^m} \frac{\gamma^{2m}}{\sqrt{g}} e^{-\oint_0^{1/\gamma^2} dt\left(\frac{1}{2}g_{\mu\nu}\dot{x}^\mu \dot{x}^\nu - i\overline{\psi}\nabla_t\psi - \frac{1}{4}R^{\rho\sigma}{}_{\mu\nu}\overline{\psi}_\rho\overline{\psi}_\sigma\psi^\mu\psi^\nu\right)},$$

$$(6.96)$$

Using heat-kernel techniques one finds:

$$\chi_V(\mathcal{E}) = \mathrm{Tr}\left[(-1)^F e^{-\frac{1}{\gamma^2}H}\right], \qquad (6.97)$$

where F is the fermion number operator. In the limit $\gamma \to 0$ only the zero-modes of H survive and therefore one must count harmonic forms with signs, which come from $(-1)^F$, leading to the result:

$$\chi(\mathcal{E}) = \sum_{k=0}^{2m}(-1)^k b_k = \chi(X), \qquad (6.98)$$

(b_k are the Betti numbers of X) in perfect agreement with our previous calculation.

Actually, owing to supersymmetry, for each non-zero energy bosonic mode there is a fermionic one with the same energy, which cancels its contribution to (6.97). Therefore the computation performed in the Hamiltonian formalism holds for any γ.

2) Topological sigma models

To further clarify the infinite-dimensional case we shall present a brief description of topological sigma models in the framework of the Mathai–Quillen formalism. The physical origin of these models is the following: there is a two-dimensional version of the twisting procedure explained in the previous chapter that can be applied to any $\mathcal{N} = 2$ supersymmetric theory. It turns out that in two dimensions there are two inequivalent ways of doing the twisting, and therefore given an $\mathcal{N} = 2$ supersymmetric field theory one obtains two different topological quantum field theories of cohomological type. These two inequivalent theories are called A and B models. When the twisting procedure is applied to the $\mathcal{N} = 2$ supersymmetric non-linear sigma model in two dimensions, one obtains the A and the B version of the topological sigma model. In this section we shall be dealing with the A model which, in its more general form, is defined in terms of a smooth, almost Hermitian manifold X, with metric $G_{\mu\nu}$ and almost complex structure $J^\mu{}_\nu$ satisfying:

$$J^\mu{}_\nu J^\nu{}_\sigma = -\delta^\mu{}_\sigma, \qquad G_{\mu\nu} J^\mu{}_\sigma J^\nu{}_\omega = G_{\sigma\omega}. \qquad (6.99)$$

In this case the configuration space $\mathcal{X}$ is the set of maps from a Riemann surface Σ to X, $x : \Sigma \to X$. The symmetry group is again trivial, and the vector bundle $\mathcal{E} \to \mathcal{X}$ has as its fiber $\mathcal{F} = \Gamma\big(T^*(\Sigma) \otimes x^*(TX)\big)^+$ over $x \in \mathcal{X}$. The superscript $^+$ means that we take the self-dual part of this vector space, i.e., if $\varrho^\mu_\alpha \in \mathcal{F}$ it satisfies the self-duality constraint $\varrho^\mu_\alpha J^\nu{}_\mu \epsilon_\beta{}^\alpha = \varrho^\nu_\beta$. The choice of section in $\mathcal{E}$ is the following:

$$s(x)^\mu_\alpha = \partial_\alpha x^\mu + J^\mu{}_\nu \epsilon_\alpha{}^\beta \partial_\beta x^\nu. \qquad (6.100)$$

Notice that this is indeed a section of $\mathcal{E}$, since it satisfies the self-duality condition $s(x)^\nu_\beta = s(x)^\mu_\alpha J^\nu{}_\mu \epsilon_\beta{}^\alpha$.

We will restrict the discussion to the simplest case in which the manifold X is Kähler. Following the general pattern, we introduce two sets of fields (x^μ, χ^μ), $(\varrho^\mu_\alpha, B^\mu_\alpha)$ for the base space $\mathcal{X}$ and the fiber $\mathcal{F}$, respectively. Here, $x^\mu(\sigma)$ is a two-dimensional field, which in this case is a map from Σ to X written in local coordinates. Following (6.40), we see that the localizing gauge fermion is given by

$$\Psi_{\text{loc}}(x, \chi, \varrho, B) = \frac{1}{2} \int_\Sigma d^2\sigma \sqrt{g} \left[\varrho^\alpha_\mu (B^\mu_\alpha + 2is^\mu_\alpha + \Gamma^\mu_{\nu\sigma} \chi^\nu \varrho^\sigma_\alpha) \right]. \qquad (6.101)$$

The model is invariant under the symmetry transformations:

$$\begin{aligned} \delta x^\mu &= \chi^\mu, & \delta \varrho^\mu_\alpha &= B^\mu_\alpha, \\ \delta \chi^\mu &= 0, & \delta B^\mu_\alpha &= 0. \end{aligned} \qquad (6.102)$$

After integrating out the auxiliary fields the action $S = \delta \Psi_{\text{loc}}$ reads:

$$\begin{aligned} S(x, \chi, \varrho) = \int_\Sigma d^2\sigma \sqrt{g} \Big(&\frac{1}{2} G_{\mu\nu} h^{\alpha\beta} \partial_\alpha x^\mu \partial_\beta x^\nu + \frac{1}{2} \epsilon^{\alpha\beta} J_{\mu\nu} \partial_\alpha x^\mu \partial_\beta x^\nu \\ &- i h^{\alpha\beta} G_{\mu\nu} \varrho^\mu_\alpha \mathcal{D}_\beta \chi^\nu - \frac{1}{8} h^{\alpha\beta} R_{\mu\nu\sigma\tau} \varrho^\mu_\alpha \varrho^\nu_\beta \chi^\sigma \chi^\tau \Big), \end{aligned} \qquad (6.103)$$

where $\mathcal{D}_\alpha\chi^\mu = \partial_\alpha\chi^\mu + \Gamma^\mu_{\nu\sigma}\partial_\alpha x^\nu\chi^\sigma$. We also have the ghost-number assignment:

$$x^\mu \longrightarrow 0, \qquad \chi^\mu \longrightarrow 1, \qquad \rho^\mu_\alpha \longrightarrow -1, \qquad B^\mu_\alpha \longrightarrow 0. \qquad (6.104)$$

Rewriting (6.100) in terms of holomorphic indices $\alpha \to (z,\bar{z})$ and $\mu \to (M, \overline{M})$, the equation for the zero locus of the section becomes:

$$\partial_\alpha x^\mu + J^\mu_{\ \nu}\epsilon_\alpha^{\ \beta}\partial_\beta x^\nu = 0 \to \partial_{\bar{z}}x^M = 0, \qquad (6.105)$$

i.e., it corresponds to the space of holomorphic instantons. The virtual dimension of the moduli space of holomorphic instantons can be obtained with the help of an index theorem. Contrary to the case of supersymmetric quantum mechanics, this dimension is generally not zero. This implies that, in general, one is forced to insert operators $\mathcal{O}$ in the integrals like (6.72) in order to obtain non-trivial results.

In general, the operators $\mathcal{O}$ are cohomology classes on $\mathcal{X}$ and are obtained from the analysis of the δ-cohomology associated to the symmetry (6.102). These operators are the observables of the cohomological field theory. In the case of topological sigma models, the highest ghost number operators turn out to be:

$$\mathcal{O}^{(0)}_A = A_{i_1,\dots,i_p}\chi^{i_1}\chi^{i_2}\cdots\chi^{i_p}, \qquad A \in \Omega^*(X), \qquad (6.106)$$

and satisfy the relation:

$$\{Q, \mathcal{O}^{(0)}_A\} = \mathcal{O}^{(0)}_{dA}, \qquad (6.107)$$

where Q denotes the generator of the symmetry δ. This relation allows us to identify the Q-cohomology classes of the highest ghost number observables with the (de Rham) cohomology classes of X. As we are dealing with a cohomological field theory, the observables satisfy topological descent equations (5.6), which in this case take the form:

$$d\mathcal{O}^{(0)}_A = \{Q, \mathcal{O}^{(1)}_A\}, \qquad d\mathcal{O}^{(1)}_A = \{Q, \mathcal{O}^{(2)}_A\}. \qquad (6.108)$$

They are easily solved:

$$\begin{aligned}
\mathcal{O}^{(1)}_A &= A_{i_1,\dots,i_p}\partial_\alpha x^{i_1}\chi^{i_2}\cdots\chi^{i_p}d\sigma^\alpha, \\
\mathcal{O}^{(2)}_A &= \frac{1}{2}A_{i_1,\dots,i_p}\partial_\alpha x^{i_1}\partial_\beta x^{i_2}\chi^{i_3}\cdots\chi^{i_p}d\sigma^\alpha \wedge d\sigma^\beta.
\end{aligned} \qquad (6.109)$$

With the help of these operators one completes the family of observables
which, as explained in the previous chapter, are labelled by homology classes
of the two-dimensional manifold Σ:

$$\int_{\gamma} \mathcal{O}_A^{(1)}, \qquad \int_{\Sigma} \mathcal{O}_A^{(2)}. \tag{6.110}$$

Topological sigma models can be generalized by including potential terms, and
the resulting theories can be also understood in the context of the Mathai–
Quillen formalism by using an equivariant extension with respect to global
abelian symmetries. We refer the reader to the references at the end of the
chapter for a detailed exposition.

6.5. The Mathai–Quillen formalism for theories with gauge symmetry

In the previous section we have considered the Mathai–Quillen formalism in the infinite-dimensional setting in order to formulate supersymmetric
quantum mechanics and topological sigma models. None of these models
involves a gauge symmetry, and the Mathai–Quillen formalism reproduces
the standard construction of these models in field-theoretic terms. However,
in theories with a gauge symmetry the Mathai–Quillen formalism has to be
modified in order to make contact with the field-theoretic construction. They
key observation, made originally by Atiyah and Jeffrey, is that one has to consider the Mathai–Quillen representative in the Cartan model of equivariant
cohomology.

In order to present this reformulation of the Mathai–Quillen formalism
let us come back to the Mathai–Quillen representative of a vector bundle $\mathcal{E}$
constructed as an associated vector bundle to a principal G-bundle P. The
universal Thom form $U(\mathcal{E})$ given in (6.28) is written in the Weil formulation
of equivariant cohomology. To obtain a Cartan model representative, one first
has to set $\theta_{ab} = 0$. After applying the Chern–Weil homomorphism one finds
the following form,

$$\Phi_C(\mathcal{E}) = (2\pi)^{-m} \mathrm{Pf}\,(K) \exp\left(-\frac{1}{2}|v|^2 - \frac{1}{2} dv_a K_{ab}^{-1} dv_b \right). \tag{6.111}$$

This form is still invariant under the action of G, but it is not a basic differential form in $\Omega(P \times V)$ since it is not horizontal. Therefore a horizontal
projection must be enforced. The result of this projection will be a basic

form $\Phi_C^h(\mathcal{E})$ (where h denotes the horizontal projection) which coincides in fact with (6.28), since the horizontal projection only applies to dv_a and gives

$$(dv_a)^h = dv_a + \theta_{ab}v_b. \tag{6.112}$$

We will now enforce the horizontal projection in a more subtle way, following Atiyah and Jeffrey. Let us assume that P is endowed with a Riemannian metric g which is G-invariant. We use this metric to define a connection on P by declaring the horizontal subspace to be the orthogonal complement to the vertical one. More explicitly, we start with the map defining fundamental vector fields on P:

$$C_p : \mathbf{g} \to T_p P, \tag{6.113}$$

and then consider the differential form on P with values in $\mathbf{g}^*$ given by:

$$\tilde{\nu}_p(Y_p, A) = g_p(C_p A, Y_p), \quad Y_p \in T_p P, \quad A \in \mathbf{g}. \tag{6.114}$$

We can then use the Killing form on $\mathbf{g}$ to obtain a one-form on P with values in $\mathbf{g}$, denoted by ν. Notice that $\nu_p = C_p^\dagger$, the adjoint of C_p, which is defined by the metric on P together with the Killing form on $\mathbf{g}$. If $R = C^\dagger C$ the connection one-form is defined by

$$\theta = R^{-1}\nu. \tag{6.115}$$

The curvature of this connection can be easily computed on horizontal vectors. In this case it is simply given by

$$K = d\theta = R^{-1}d\nu, \tag{6.116}$$

as the other terms vanish on the horizontal subspace. Let us now write (6.111) as a Grassmann integral:

$$\Phi_C(\mathcal{E}) = (2\pi)^{-m} \int d\chi\, e^{-\frac{1}{2}|v|^2 + \frac{1}{2}\chi_a K_{ab}\chi_b + i\chi_a dv_a}. \tag{6.117}$$

Since at the end of the process we want to make a horizontal projection of this form, we can write $K = d\theta = R^{-1}d\nu$. We can now introduce Lie algebra variables λ, ϕ and use a Fourier transform formula to obtain the expression:

$$\Phi_C(\mathcal{E}) = (2\pi)^{-d-m} \int d\chi d\phi d\lambda \det R \, \exp\Big(-\frac{1}{2}|v|^2 + \frac{1}{2}\chi_a \phi_{ab}\chi_b$$
$$+ i\chi_a dv_a + i\langle d\nu, \lambda\rangle_g - i\langle \phi, R\lambda\rangle_g\Big), \tag{6.118}$$

where $\langle\, ,\, \rangle_g$ denotes the Killing form of $\mathbf{g}$, and $d = \dim(G)$. Notice that in this expression the integration over λ gives a δ-function constraining ϕ to be K. We now enforce the horizontal projection of $\Phi_C(\mathcal{E})$ by multiplying it by the normalized invariant volume form $\mathcal{D}g$ along the G-orbits. One can show that

$$(\det R)\,\mathcal{D}g = \int d\eta \exp i(\langle\eta, \nu\rangle_g), \qquad (6.119)$$

where η is a fermionic Lie algebra variable. Putting everything together we obtain a representative for the horizontal projection:

$$\Phi(\mathcal{E})_C^h = (2\pi)^{-d-m} \int d\eta d\chi d\phi d\lambda \exp\Big(-\frac{1}{2}|v|^2 + \frac{1}{2}\chi_a\phi_{ab}\chi_b + i\chi_a dv_a$$
$$+ i\langle dv, \lambda\rangle_g - i\langle\phi, R\lambda\rangle_g + i\langle\eta, \nu\rangle_g \Big). \qquad (6.120)$$

This is now a basic form in $\Omega(P \times V)$ and descends to the Thom form of $\mathcal{E}$.

We can also introduce a BRST structure for the action involved in the exponent of (6.120). First, we introduce auxiliary fields B_a with the meaning of a basis of differential forms for the fiber. The natural BRST operator δ is again the de Rham d operator, but we must take into account that we have to enforce the horizontal projection of dv_a given in (6.112). This leads to the following modified transformation:

$$\delta\chi_a = B_a, \qquad \delta B_a = \phi_{ab}\chi_b. \qquad (6.121)$$

Since we must also take into account the horizontal projection of differential forms on P, we conclude that δ is nothing but the the equivariant exterior derivative:

$$\delta = d - \iota(C\phi). \qquad (6.122)$$

Notice that ϕ is an element of the Lie algebra $\mathbf{g}$, and therefore $C\phi$ is a fundamental vector field on P. The transformation (6.121) can be also interpreted as the equivariant derivative acting on the fiber V, after taking into account that the group G acts as $v \to g^{-1}v$ on $v \in V$.

Owing to the horizontal projection, we need, in fact, two gauge fermions in order to reconstruct (6.120): a localizing gauge fermion

$$\Psi_{\mathrm{loc}} = \chi_a(iv_a + \frac{1}{2}B_a); \qquad (6.123)$$

and a 'projecting' one

$$\Psi_{\mathrm{proj}} = i\langle\lambda, \nu\rangle. \qquad (6.124)$$

The BRST transformation acts on the Lie algebra elements as

$$\delta\lambda = \eta, \qquad \delta\eta = -[\phi, \lambda]. \tag{6.125}$$

Using (6.125), (6.121), and (6.122) as BRST operators acting on the gauge fermion $\Psi = \Psi_{\text{loc}} + \Psi_{\text{proj}}$, one recovers the action in (6.120). Notice that since the natural δ symmetry is the equivariant derivative, the observables associated with the model will live in the equivariant cohomology with respect to the gauge symmetry.

6.6. Donaldson–Witten theory in the Mathai–Quillen formalism

Donaldson–Witten theory was first analyzed from the viewpoint of the Mathai–Quillen formalism by Atiyah and Jeffrey, following the general formalism presented in the previous section. The basic geometric framework for the moduli problem is as in Chapter 2: a compact oriented four-dimensional manifold X endowed with a metric $g_{\mu\nu}$. Over this manifold X we have a principal G-bundle $P \to X$, together with an associated vector bundle E. According to our previous general discussion on the Mathai Quillen formalism, to construct a topological quantum field theory associated to Donaldson theory we have to specify fields, symmetries and equations. Given the basic geometrical framework, our insight from the considerations in previous chapters facilitates the choices which have to be made.

The configuration space of the theory $\mathcal{X}$ is just the space $\mathcal{A}$ of G-connections on E, and the symmetry $\mathcal{G}$ is the group of gauge transformations on $\mathcal{A}$. The bundle $\mathcal{E}$ is defined as $\mathcal{A} \times_{\mathcal{G}} \mathcal{F}$, where $\mathcal{F}$ is the vector space of self-dual two-forms on X with values in $\mathbf{g}_E$, $\Omega^{2,+}(X, \mathbf{g}_E)$. The map $s : \mathcal{A} \to \mathcal{F}$ given by

$$s(A) = F^+(A) \tag{6.126}$$

descends to a section $s : \mathcal{A}/\mathcal{G} \to \mathcal{E}$, as we already saw in (2.23).

In order to complete the construction we must specify the field content of the theory. As should be clear from the previous sections, there is a pair of fields associated with the base space $\mathcal{A}$, (A_μ, ψ_μ), and two other fields associated with the fiber $\mathcal{F}$, $(\chi_{\mu\nu}, G_{\mu\nu})$. Since we are going to use the Mathai–Quillen representative constructed in the previous section, we also need the $\mathbf{g}_E$-valued fields λ, ϕ, and η. In total we have the following set of fields:

$$\chi_{\mu\nu}, \; G_{\mu\nu} \in \Omega^{2,+}(X, \mathbf{g}_E), \qquad \psi_\mu \in \Omega^1(X, \mathbf{g}_E), \qquad \eta, \; \lambda, \; \phi \in \Omega^0(X, \mathbf{g}_E), \tag{6.127}$$

which is indeed the field content of Donaldson–Witten theory. The ghost number assignment is

$$A_\mu \longrightarrow 0, \qquad \chi_{\mu\nu} \longrightarrow -1, \qquad G_{\mu\nu} \longrightarrow 0, \qquad \psi_\mu \longrightarrow 1,$$
$$\eta \longrightarrow -1, \qquad \lambda \longrightarrow -2, \qquad \phi \longrightarrow 2, \tag{6.128}$$

and agrees with the ghost number inherited from the $U(1)_R$ symmetry of the $\mathcal{N} = 2$ supersymmetric theory. The field ψ_μ lives in the (co)tangent space to the field space and is to be understood as providing a basis for differential forms on $\mathcal{A}$, whereas the scalar bosonic field ϕ —or rather its expectation value $\langle \phi \rangle$— plays the role of the curvature two-form of the principal $\mathcal{G}$-bundle $\mathcal{A} \to \mathcal{A}/\mathcal{G}$.

The scalar symmetry δ which characterizes the theory is the equivariant derivative with parameter ϕ, and has the form:

$$\begin{aligned} \delta A_\mu &= \psi_\mu, \\ \delta \chi_{\mu\nu} &= G_{\mu\nu}, \\ \delta \lambda &= \eta, \\ \delta \phi &= 0. \end{aligned} \qquad \begin{aligned} \delta \psi &= \nabla_A \phi, \\ \delta G_{\mu\nu} &= i[\chi_{\mu\nu}, \phi], \\ \delta \eta &= i[\lambda, \phi], \end{aligned} \tag{6.129}$$

Notice that since δ is an equivariant derivative δ^2 is a gauge transformation with gauge parameter ϕ, and in order to obtain the observables one is led to study the $\mathcal{G}$-equivariant cohomology of δ. These transformations are equivalent to those obtained in (5.29) after the redefinitions:

$$\phi \longrightarrow 2\sqrt{2}\phi, \qquad \lambda \longrightarrow -\frac{i}{2\sqrt{2}}\phi^\dagger, \qquad G_{\mu\nu} \longrightarrow i(F_{\mu\nu}^+ - D_{\mu\nu}). \tag{6.130}$$

The action of the theory is δ-exact and can be written in terms of the localizing and projecting gauge fermions:

$$\Psi_{\rm loc} = \int_X d^4x \sqrt{g}\, {\rm Tr} \left[2\chi_{\mu\nu}(F^{+\mu\nu} - \frac{1}{2}G^{\mu\nu}) \right],$$
$$\Psi_{\rm proj} = \int_X d^4x \sqrt{g}\, {\rm Tr} \left[i\lambda \nabla_\mu \psi^\mu \right], \tag{6.131}$$

where we have taken into account that $\nu = C^\dagger = \nabla_A^\dagger$, as explained in (2.38). After integrating out the auxiliary fields the action reads:

$$\delta(\Psi_{\rm loc} + \Psi_{\rm proj}) \to \int_X d^4x \sqrt{g}\, {\rm Tr} \Big(F^{+2} - i\chi^{\mu\nu}\nabla_\mu \psi_\nu + i\eta \nabla_\mu \psi^\mu$$
$$+ \frac{1}{4}\phi\{\chi_{\mu\nu}, \chi^{\mu\nu}\} + \frac{i}{4}\lambda\{\psi_\mu, \psi^\mu\} - \lambda \nabla_\mu \nabla^\mu \phi \Big). \tag{6.132}$$

This action differs from the one obtained by the twisting procedure in (5.28) by a δ-exact term. Indeed, comparing (5.31) and (6.131), after taking into account the redefinitions (6.130), one obtains that their difference is a term involving $\text{Tr}\,\eta[\phi, \lambda]$. Since terms of this type in the action are irrelevant we conclude that the construction of the Euler class of the infinite-dimensional bundle $\mathcal{E}$ by using the Mathai–Quillen formalism leads to Donaldson–Witten theory. We can then rederive in a geometric way the property of the Donaldson–Witten functional integral reducing to a finite-dimensional integral over the moduli space of ASD instantons (the zero locus of the section (6.126)).

6.7. Abelian monopoles in the Mathai–Quillen formalism

In this section we will consider abelian monopoles in the Mathai–Quillen formalism. We will recover the model obtained in the previous chapter from the approach based on twisted $\mathcal{N} = 2$ supersymmetry.

As in the case of Donaldson–Witten theory we will start identifying fields, symmetries and equations. The basic geometric framework is as in Chapter 3. Let X be a Riemannian four-manifold, and let L be the determinant line bundle of a Spin_c connection. The configuration space for the moduli problem is $\mathcal{X} = \mathcal{A} \times \Gamma(X, S^+ \otimes L^{\frac{1}{2}})$, and the symmetry $\mathcal{G}$ is the group of gauge transformations acting on $\mathcal{X}$ by (3.4). The vector bundle $\mathcal{E}$ is $\mathcal{X} \times_{\mathcal{G}} \mathcal{F}$, where $\mathcal{F} = \Omega^{2,+}(X) \oplus \Gamma(X, S^- \otimes L^{\frac{1}{2}})$. The map $s : \mathcal{X} \to \mathcal{F}$ given in (3.8) descends to a section of $\mathcal{E}$, and the moduli space $\mathcal{M}$ is the space of solutions to the Seiberg–Witten equations

$$F^+_{\dot\alpha\dot\beta} + 4i\overline{M}_{(\dot\alpha} M_{\dot\beta)} = 0,$$
$$D_L^{\alpha\dot\alpha} M_{\dot\alpha} = 0, \tag{6.133}$$

modulo gauge transformations. Using this section we can construct the corresponding topological field theory in the Mathai–Quillen formalism. In addition to the fields in Donaldson–Witten theory, we also have to add two pairs of fields, one pair,

$$M_{\dot\alpha}, \mu_{\dot\alpha} \in \Gamma(S^+ \otimes L^{\frac{1}{2}}), \tag{6.134}$$

corresponding to the base manifold $\mathcal{X}$, and another pair,

$$\nu_\alpha, h_\alpha \in \Gamma(S^- \otimes L^{\frac{1}{2}}), \tag{6.135}$$

corresponding to the fiber $\mathcal{F}$, together with their corresponding complex conjugates. The ghost number assignments for these new fields are those prescribed by our general rule (6.74):

$$M_{\dot\alpha} \longrightarrow 0, \qquad \mu_{\dot\alpha} \longrightarrow 1, \qquad \nu_\alpha \longrightarrow -1, \qquad h_\alpha \longrightarrow 0. \qquad (6.136)$$

On the other hand, the scalar symmetry which characterizes the theory is now enlarged in the following way:

$$\begin{aligned} \delta M_{\dot\alpha} &= \mu_{\dot\alpha}, & \delta\mu_{\dot\alpha} &= -i\phi M_{\dot\alpha}, \\ \delta v_\alpha &= h_\alpha, & \delta h_\alpha &= -i\phi v_\alpha. \end{aligned} \qquad (6.137)$$

The gauge fermion of the theory takes the following form:

$$\begin{aligned} \Psi_{\text{loc}} &= \int_X d^4x\, \sqrt{g}\Big(\frac{i}{2}\chi^{\dot\alpha\dot\beta}\big(\frac{1}{\sqrt2}(F^+_{\dot\alpha\dot\beta} + 4i\overline{M}_{(\dot\alpha}M_{\dot\beta)}) - \frac{i}{4}G_{\dot\alpha\dot\beta}\big) \\ &\quad - \frac{i}{2}(\overline{v}_\alpha\nabla^{\alpha\dot\alpha}M_{\dot\alpha} - \overline{M}^{\dot\alpha}\nabla_{\alpha\dot\alpha}v^\alpha) - \frac{1}{8}(\overline{v}_\alpha h^\alpha + \overline{h}_\alpha v^\alpha)\Big), \\ \Psi_{\text{proj}} &= -\frac{1}{2}\int_X d^4x\, \sqrt{g}\Big(i\text{Tr}(\lambda \wedge *\nabla^*_A\psi) + e(\overline{\mu}^\alpha\lambda M_\alpha - \overline{M}^\alpha\lambda\mu_\alpha)\Big). \end{aligned} \qquad (6.138)$$

Notice that after the redefinitions (6.130) this agrees with (5.57) for $m = 0$ up to an overall factor and the δ-exact term $\text{Tr}\,\eta[\phi,\lambda]$. After integrating out the auxiliary fields, the action reads:

$$\begin{aligned} S_{\text{AM}} &= \int_X d^4x\, \sqrt{g}\Big(g^{\mu\nu}\nabla_\mu\overline{M}^{\dot\alpha}\nabla_\nu M_{\dot\alpha} + \frac{1}{4}R\overline{M}^{\dot\alpha}M_{\dot\alpha} + \frac{1}{2}F^{+\dot\alpha\dot\beta}F^+_{\dot\alpha\dot\beta} \\ &\quad - \frac{1}{8}\overline{M}^{(\dot\alpha}M^{\dot\beta)}\overline{M}_{(\dot\alpha}M_{\dot\beta)}\Big) \\ &\quad + \int_X \text{Tr}\Big(\frac{i}{2}\eta \wedge *\nabla^*_A\psi + \frac{i}{2\sqrt2}\chi^{\dot\alpha\dot\beta}(p^+(\nabla_A\psi))_{\dot\alpha\dot\beta} + \frac{i}{2}\lambda \wedge *\nabla^*_A\nabla_A\phi\Big) \\ &\quad + \int_X d^4x\, \sqrt{g}\Big(i\overline{M}^{\dot\alpha}\phi\lambda M_{\dot\alpha} - \frac{1}{\sqrt2}(\overline{M}_{\dot\alpha}\chi^{\dot\alpha\dot\beta}\mu_{\dot\beta} - \overline{\mu}_{\dot\alpha}\chi^{\dot\alpha\dot\beta}M_{\dot\beta}) \\ &\quad - \frac{i}{2}(\overline{v}_\alpha\nabla^{\alpha\dot\alpha}\mu_{\dot\alpha} + \overline{\mu}^{\dot\alpha}\nabla_{\alpha\dot\alpha}v^\alpha) \\ &\quad + \frac{1}{2}(\overline{M}^{\dot\alpha}\psi_{\alpha\dot\alpha}v^\alpha + \overline{v}_\alpha\psi^{\alpha\dot\alpha}M_{\dot\alpha}) \\ &\quad - \frac{1}{2}(\overline{\mu}^{\dot\alpha}\eta M_{\dot\alpha} + \overline{M}^{\dot\alpha}\eta\mu_{\dot\alpha}) + \frac{i}{4}\overline{v}^\alpha\phi v_\alpha - \overline{\mu}^{\dot\alpha}\lambda\mu_{\dot\alpha}\Big). \end{aligned} \qquad (6.139)$$

Bibliographical notes

- The definitive reference on cohomological field theories from the point of view of the Mathai–Quillen formalism is [9]. The geometric background on bundles and the Thom class can be found in [65] and [66]. The formula (6.17) for non-generic sections is derived in [67]. The original Mathai–Quillen formalism was introduced in [68]. It was first implemented in the context of TQFT by Atiyah and Jeffrey in [69]. Excellent reviews of its applications in TQFT can be found in the works [70] and [71].

- The BRST context originated after the pioneering works [72] and [73]. A review on BRST can be found in [74].

- Supersymmetric quantum mechanics was first considered from a topological point of view in [49]. Further important work on the subject, including the application of heat kernel techniques, was carried out in [75], [76] and [48]. The formulation of supersymmetric quantum mechanics and topological sigma models in the context of the Mathai–Quillen formalism can be found in [9].

- Topological sigma models where first introduced by Witten in [47], where the so called A-model was constructed. B-models were first described in [77] and [78]. Supersymmetric sigma models with potentials were constructed in [79], and their topological version was introduced in [80]. To incorporate this model in the framework of the Mathai–Quillen formalism one has to consider an equivariant extension with respect to global abelian symmetries. This extension has been worked out in [81], where the example of topological sigma models with potentials is analyzed in detail.

- The abelian monopole equations were analysed in the context of the Mathai–Quillen formalism in [55]. The generalization to the non-abelian case was studied in the same formalism in [56] and [81].

Chapter 7

The Seiberg–Witten Solution of $\mathcal{N} = 2$ SUSY Yang–Mills Theory

The Seiberg–Witten solution of $SU(2)$, $\mathcal{N} = 2$ supersymmetric gauge theories obtained by Seiberg and Witten in 1994 provides the exact low energy effective action (LEEA) of $\mathcal{N} = 2$ supersymmetric Yang–Mills theory up to two derivatives. In this book we can not make a detailed analysis of this remarkable achievement, and we will just provide some starting points and the final results, which will be needed in the topological applications of the remaining chapters. There are excellent reviews which the reader can consult for further information, and some of them are addressed in the bibliographical notes at the end of this chapter. We will start by reviewing some semi-classical aspects of the theory that motivate the ingredients for the exact solution. $\mathcal{N} = 2$ supersymmetry reduces the problem of finding the LEEA to the problem of finding a single holomorphic function, the so called prepotential. The duality properties of the prepotential are reviewed in the second section. Since the Seiberg–Witten solution is written in terms of an auxiliary elliptic curve, we will review in some detail some of the basic aspects of elliptic curves and the theory of elliptic functions. With this machinery we shall present the exact solution in section 4 and explore some of its properties. Finally, we shall re-express some aspects of the solution in terms of modular forms, something that will be very useful for analysing the u-plane integral.

7.1. Low energy effective action: semi-classical aspects

1) Classical potential and moduli space

In order to analyse the $\mathcal{N} = 2$ supersymmetric $SU(2)$ theory the first step is to perform a classical analysis. The classical action (4.67) contains the scalar potential

$$V = \frac{1}{2g^2} \text{Tr} \, [\phi^\dagger, \phi]^2 \,. \tag{7.1}$$

The classical vacua are determined by $V = 0$, *i.e.*, $[\phi, \phi^\dagger] = 0$. This means that ϕ is gauge equivalent to an element in the Cartan subalgebra, and

$$\phi = \begin{pmatrix} a & 0 \\ 0 & -a \end{pmatrix}. \tag{7.2}$$

When ϕ is of this form the gauge symmetry is broken to $U(1)$. The vacuum expectation value a gives mass to the $W^\pm$ bosons in the usual way. The final outcome of this process is that, of the three $\mathcal{N} = 2$ supersymmetric vector superfields of the $SU(2)$ theory two of them (the $\mathcal{N} = 2$ supersymmetric multiplets of the $W^\pm$ bosons) obtain a mass, whilst one of them remains massless. Classically, the only light degree of freedom is a $U(1)$ $\mathcal{N} = 2$ supersymmetric multiplet.

Notice that different vacuum values of ϕ correspond to different physical theories, and therefore the complex number a parametrizes the space of physically inequivalent vacua. This space is known as the *classical moduli space* of the theory, and in this case it has complex dimension one. However, this parametrization of the moduli space is not gauge-invariant. Although we have used the gauge symmetry to put ϕ in the form (7.2), there is still some residual gauge invariance: a Weyl reflection will take $a \to -a$. It is more convenient to parametrize the moduli space by the vacuum expectation value of a gauge-invariant operator,

$$u = \langle \mathrm{Tr}\,(\phi^2) \rangle, \tag{7.3}$$

which defines the so called u-plane. Classically $u = 2a^2$, but as we will see this becomes corrected in the full quantum theory.

2) Constraints from $\mathcal{N} = 2$ supersymmetry

Our main goal is to find the effective quantum theory for the massless degree of freedom, which classically is a $U(1)$ $\mathcal{N} = 2$ supersymmetric multiplet $\mathcal{A}$. In $\mathcal{N} = 1$ language, this contains an $\mathcal{N} = 1$ chiral multiplet

$$A = a - \sqrt{2}\theta\lambda_1 + \theta^2 F, \tag{7.4}$$

and an $\mathcal{N} = 1$ vector multiplet

$$V = -2\theta\sigma^\mu\bar{\theta}A_\mu - 2i\bar{\theta}^2\theta\lambda_2 + 2i\theta^2\bar{\theta}\bar{\lambda}^2 + \theta^2\bar{\theta}^2 D. \tag{7.5}$$

Notice that we have denoted by a both the lowest component of the $U(1)$ chiral multiplet, in (7.4), and its vacuum expectation value in (7.2). The choices (7.4) and (7.5) correspond to (4.18) and (4.36) after some obvious redefinitions.

In principle, the LEEA should be obtained by integrating out the massive degrees of freedom, *i.e.*, by performing the functional integral over all the modes of the massive multiplets. Since this procedure gives in general a non-local action one has to expand in derivatives to obtain a local effective theory. The expansion is then in powers of p/Λ, where p is the momentum and Λ is the typical scale in the theory (in this case, as we shall see, since the theory is asymptotically free Λ is the dynamically generated scale).

Of course, in real life it is extremely hard to integrate out the massive degrees of freedom explicitly, and one obtains the structure of low energy effective actions by symmetry considerations (as, for example, in chiral perturbation theory). One of the big advantages of $\mathcal{N} = 2$ supersymmetry is that the leading terms in the derivative expansion (the terms that contain at most two derivatives and four fermions) can be obtained from a single holomorphic function $\mathcal{F}(a)$ called the prepotential. As explained in section 5.4 this result is obtained after considering the most general $\mathcal{N} = 2$ supersymmetric action involving a single vector multiplet. The form of this action was displayed in (4.70). Adapting it to the fields involved in the low energy effective Lagrangian one finds:

$$\mathcal{L}_{\text{eff}} = \frac{1}{4\pi} \operatorname{Im} \left[\int d^4\theta \, \frac{\partial \mathcal{F}(A)}{\partial A} \overline{A} + \int d^2\theta \, \frac{1}{2} \frac{\partial^2 \mathcal{F}}{\partial A^2} W^\alpha W_\alpha \right], \qquad (7.6)$$

W^α being defined in (4.37). In terms of $\mathcal{F}(a)$ the Kähler potential in (4.28) is given by

$$K = \operatorname{Im} \left(\overline{A} \frac{\partial \mathcal{F}(A)}{\partial A} \right). \qquad (7.7)$$

If a denotes the scalar component of the chiral superfield A, then the metric on the space of fields (or equivalently on the moduli space of vacua), is given by

$$ds^2 = g_{a\bar{a}} \, da d\bar{a} = \operatorname{Im}\tau \, da d\bar{a}, \qquad (7.8)$$

where

$$\tau(a) = \frac{\partial^2 \mathcal{F}}{\partial a^2}. \qquad (7.9)$$

$\mathcal{N} = 2$ supersymmetry then implies taht the metric is given by the gauge coupling τ.

In component fields the above Lagrangian becomes:

$$
\begin{aligned}
& -\frac{1}{16\pi}\mathrm{Im}\left[\tau(F_{\mu\nu}^2 - iF_{\mu\nu}*F_{\mu\nu})\right] - \frac{1}{4\pi}\left[\tau\partial_\mu a\partial^\mu\overline{a} + i\tau\lambda^v\sigma^\mu\partial_\mu\overline{\lambda}_v\right] \\
& -\frac{\sqrt{2}}{16\pi}\mathrm{Im}\left[\frac{d\tau}{da}\lambda^v\sigma^{\mu\nu}\lambda_v F_{\mu\nu}\right] - \frac{\sqrt{2}}{16\pi}\mathrm{Im}\left[\frac{d\tau}{da}\lambda^v\lambda^w D_{vw}\right] \\
& +\frac{1}{16\pi}\mathrm{Im}\tau D_{vw}D^{vw} + \frac{1}{48\pi}\mathrm{Im}\left[\frac{d^2\tau}{da^2}(\lambda^v\lambda^w)(\lambda_v\lambda_w)\right].
\end{aligned}
\tag{7.10}
$$

Classically the low energy $U(1)$ theory describing the massless photon of $\mathcal{N} = 2$ supersymmetric Yang–Mills theory with gauge group $SU(2)$ is given by the prepotential $\mathcal{F}(a) = a^2/2$. In terms of the u variable, the metric in moduli space reads

$$
ds^2 = \frac{du d\overline{u}}{8|u|},
\tag{7.11}
$$

which is singular at $u = 0$. This has a simple physical interpretation: at $u = 0$ the $W^\pm$ bosons are, in fact, massless and the $SU(2)$ symmetry is restored. This is a very simple example of a general phenomenon which will be realized later in a much more non-trivial way: whenever a particle in the spectrum becomes massless (in this case the $W^\pm$ bosons) the low energy theory obtained from integrating out the massive degrees of freedom becomes singular.

3) The prepotential to one-loop

The classical analysis and the constraints of $\mathcal{N} = 2$ supersymmetry give the general picture of the LEEA reviewed above. In order to make further progress we have to investigate the quantum dynamics. The first thing we can do is to evaluate the perturbative one-loop correction to $\mathcal{F}$. The effective coupling of the theory can be computed as follows. At energies larger than a the masses of the $W^\pm$ bosons are negligible, and we can use the β-function of the $SU(2)$ theory. The β-function for a $SU(N_c)$ gauge theory with N_f Weyl fermions in the representations R_f and N_s complex scalar fields in the representations R_s is given by

$$
\beta(g) \equiv \mu\frac{dg}{d\mu} = -\frac{g^3}{48\pi^2}\left(11N_c - 2\sum_f C_f - \frac{1}{4}\sum_s C_s\right),
\tag{7.12}
$$

where $C_{f,s}$ are the Casimirs of the representations $R_{f,s}$, normalized in such a way that the Casimir in the adjoint is given by N_c. Since pure $\mathcal{N} = 2$

supersymmetric $SU(2)$ Yang–Mills contains one complex scalar and two Weyl fermions, both in the adjoint representation, the β-function is given by

$$\beta(g) = -\frac{1}{4\pi^2}g^3. \tag{7.13}$$

The $SU(2)$ theory is then asymptotically free, and generates a scale Λ. Below the scale a the $W^\pm$ bosons decouple, and we are left with the effective $U(1)$ theory described by the prepotential $\mathcal{F}(a)$. The coupling constant of this theory is obtained by matching the running coupling constant of the $SU(2)$ theory at the scale $\mu = a$. If $a \gg \Lambda$ we are at energies much bigger than the scale of the theory and asymptotic freedom tells us that perturbative computations are reliable. Using standard perturbation theory and the holomorphy of τ one can obtain the effective coupling constant at one loop:

$$\tau(a) = \frac{2i}{\pi}\log\left(\frac{a}{\Lambda}\right). \tag{7.14}$$

If we now integrate (7.9) we find

$$\mathcal{F}_{1-\text{loop}}(a) = \frac{i}{2\pi}a^2\ln\frac{a^2}{\Lambda^2}. \tag{7.15}$$

It is known that owing to $\mathcal{N} = 2$ supersymmetry the above one-loop expression for the prepotential does not receive higher order perturbative corrections, and (7.15) gives the full perturbative prepotential. It is customary to include the classical piece of the prepotential $a^2/2$ in the one-loop piece by redefining Λ. We will refer to the regime $u \gg \Lambda^2$ as the *semi-classical regime*.

4) Instanton effects

Since we are dealing with a non-abelian gauge theory, we should expect non-perturbative effects owed to instantons. The first possible effect of instanton backgrounds is to break some of the classical symmetries, as in the $U(1)$ problem of quantum cromodynamics. Classically, our theory has the full global $SU(2)_R \times U(1)_R$ as a symmetry group. However, in an instanton background $U(1)_R$ is broken to a discrete subgroup. This can be easily understood as follows: by the index theorem, in the presence of an $SU(N)$ instanton there are $2N$ zero-modes for each fermion λ_α in the adjoint representation, and since there is an $SU(2)_R$ doublet of fermions, there are in total $4N$ zero-modes. The measure of the functional integral

$$\mathcal{D}\overline{\lambda}_{1\dot\alpha}\mathcal{D}\lambda_{1\alpha}\mathcal{D}\overline{\lambda}_{2\dot\alpha}\mathcal{D}\lambda_{2\alpha} \tag{7.16}$$

is not invariant under the $U(1)_R$ symmetry (4.68): by the usual argument à la Fujikawa, it picks a factor

$$e^{2i\varphi\,\mathrm{index}(D)} = e^{-4iN\varphi}, \tag{7.17}$$

where D is the Dirac operator coupled to the adjoint bundle. Therefore the measure is only invariant under the discrete group $\mathbf{Z}_{4N}$. The surviving discrete R-symmetry is broken spontaneously by the Higgs vacuum expectation value. The field ϕ has charge 2 under $\mathbf{Z}_{4N}$ and transforms into $e^{\pi i/N}\phi$. Therefore, for the $SU(2)$ theory, if the vacuum is characterized by non-zero ϕ, then $\mathbf{Z}_8$ is broken down to $\mathbf{Z}_4$. The spontaneously broken symmetry has a non-trivial action on the moduli space, and it acts as $\mathbf{Z}_2$ on the u-plane: $u \to -u$.

Instantons also give non-perturbative corrections to the prepotential. The structure of these corrections can be determined as follows: first, a correction to $\mathcal{F}$ coming from a configuration of instanton number k should be proportional to the k-instanton factor $\exp(-8\pi^2 k/g^2)$ (since the prepotential is a holomorphic function, it cannot receive corrections from anti-instanton configurations). Using the explicit expression for the β-function of the theory (7.13) the k-instanton factor can be written as

$$e^{-8\pi^2 k/g^2} = \left(\frac{\Lambda}{a}\right)^{4k}. \tag{7.18}$$

To make further progress we notice that although the R-symmetry is broken by instanton effects, one can formally restore it by assigning charge 2 to the dynamically generated scale Λ. In this way the instanton factor (7.18) is neutral. The prepotential has charge 4, and this finally implies that the k-instanton correction should also be proportional to a^2. Putting these together the prepotential including generic non-perturbative corrections can be written as

$$\mathcal{F} = \frac{i}{2\pi}a^2\ln\frac{a^2}{\Lambda^2} + \sum_{k=1}^{\infty} \mathcal{F}_k\left(\frac{\Lambda}{a}\right)^{4k} a^2. \tag{7.19}$$

The coefficients $\mathcal{F}_k$ are constants (this is because in a supersymmetric theory, instantons contribute to the functional integral only through zero-modes). It turns out that the $\mathcal{F}_k$ are all non-zero. Notice that finding the exact LEEA is equivalent to computing $\mathcal{F}_k$ for all k. The Seiberg–Witten exact solution will give an explicit procedure to determine these coefficients.

5) BPS states

Spontaneously broken $SU(N)$ gauge theories with scalar fields in the adjoint representation contain 't Hooft–Polyakov monopoles and dyons in their semi-classical spectrum. These are solitons, *i.e.*, time-independent, finite energy solutions of the classical equations of motion, and they are characterized by the value (7.2) of the Higgs field at spatial infinity. A dyon of electric charge n_e and magnetic charge n_m has a mass which is given semi-classically by

$$M = \sqrt{2}|a(n_e + \tau n_m)|. \qquad (7.20)$$

The spectrum of $\mathcal{N} = 2$ supersymmetric $SU(2)$ Yang–Mills theory has a monopole of charge $n_m = 1$ and dyons of charge $(n_e, n_m) = (n, 1)$. Quantization of these solitons leads to states which preserve half of the supersymmetries and arrange themselves into an $\mathcal{N} = 2$ supersymmetric massive hypermultiplet, with central charge given semi-classically by

$$Z = a(n_e + \tau_{\text{cl}} n_m). \qquad (7.21)$$

These states satisfy the Bogomolnyi bound $M = \sqrt{2}|Z|$ and are called BPS states. As we shall see, (7.21) becomes corrected in the full quantum theory.

*7.2. Sl(2, **Z**) duality of the effective action*

One of the most important aspects of the LEEA described by an $\mathcal{N} = 2$ supersymmetric prepotential is that one can perform an Sl$(2, \mathbf{Z})$ transformation to obtain another description of the same low energy theory. These different descriptions involve different parametrizations of the quantum moduli space. Depending on the region we are looking at, some parametrizations will be more appropriate than others, and this will play a crucial role in the rest of this book.

The duality transformations of the $\mathcal{N} = 2$ supersymmetric LEEA are, in fact, a generalization of the usual duality of abelian Maxwell theory. Let us then start with the Maxwell action in Minkowski space, with conventions $(F_{\mu\nu})^2 = -(\widetilde{F}_{\mu\nu})^2$ and $\widetilde{\widetilde{F}} = -F$:

$$\frac{1}{32\pi} \text{Im} \int \tau(a)(F + i\widetilde{F})^2 = \frac{1}{16\pi} \text{Im} \int \tau(a)(F^2 + i\widetilde{F}F). \qquad (7.22)$$

Usually we take the gauge connection A_μ as the basic field, with $F = dA$, and then it follows that $dF = 0$ (the Bianchi identity). But we can regard

F as an independent field and implement the Bianchi identity by introducing a Lagrange multiplier vector field V_D. To fix the Lagrange multiplier term, $U(1) \subset SU(2)$ is normalized such that all $SO(3)$ fields have integer charges. With this convention a magnetic monopole satisfies $\epsilon^{0\mu\nu\rho} \partial_\mu F_{\nu\rho} = 8\pi\delta^{(3)}(x)$. The Lagrange multiplier term can now be constructed by coupling V_D to a monopole:

$$\frac{1}{8\pi} \int V_{D\mu}\epsilon^{\mu\nu\rho\sigma}\partial_\nu F_{\rho\sigma} = \frac{1}{8\pi} \int \widetilde{F}_D F = \frac{1}{16\pi} \operatorname{Re} \int (\widetilde{F}_D - iF_D)(F + i\widetilde{F})\,,$$
(7.23)

with $F_{D\mu\nu} = \partial_\mu V_{D\nu} - \partial_\nu V_{D\mu}$. Adding this to the gauge field action and integrating over F we obtain the dual theory

$$\frac{1}{32\pi} \operatorname{Im} \int \left(-\frac{1}{\tau}\right) (F_D + i\widetilde{F}_D)^2 = \frac{1}{16\pi} \operatorname{Im} \int \left(-\frac{1}{\tau}\right) (F_D^2 + i\widetilde{F}_D F_D)\,. \quad (7.24)$$

We want to show that something similar happens in the effective theory described by (7.6). We can, in fact, perform a generalization in $\mathcal{N} = 1$ superspace of the dualization procedure which we have presented. Let us consider first the second term in (7.6):

$$\frac{1}{8\pi} \operatorname{Im} \int d^2\theta \tau(A) W^2\,. \quad (7.25)$$

The Bianchi identity is now replaced by $\operatorname{Im}\mathcal{D}W = 0$. This can be implemented by introducing a vector superfield V_D and the corresponding Lagrange multiplier term becomes

$$\frac{1}{4\pi} \operatorname{Im} \int d^4x d^4\theta\, V_D \mathcal{D}W = \frac{1}{4\pi} \operatorname{Re} \int d^4x d^4\theta\, i\mathcal{D}V_D W$$
$$= -\frac{1}{4\pi} \operatorname{Im} \int d^4x d^2\theta\, W_D W\,. \quad (7.26)$$

Adding this to the action and integrating out W, we obtain the dual action

$$\frac{1}{8\pi} \operatorname{Im} \int d^2\theta \left(-\frac{1}{\tau(A)}W_D^2\right)\,. \quad (7.27)$$

The conclusion of this analysis is the following: the effect of the duality transformation is to replace a gauge field which couples to electric charges by a dual gauge field which couples to magnetic charges, and at the same time the gauge coupling is transformed as

$$\tau \to \tau_D = -\frac{1}{\tau}\,. \quad (7.28)$$

This is the famous electric–magnetic duality of $U(1)$ gauge theory. Notice that it transforms weak coupling (small g) in strong coupling (large g), and vice versa. Another symmetry of the theory comes from the periodicity of the Θ angle: the theory is invariant under the shift $\Theta \to \Theta + 2\pi$. Since this corresponds to $\tau \to \tau + 1$, we should consider the group $\mathrm{Sl}(2, \mathbf{Z})$ generated by these two transformations:

$$\mathrm{Sl}(2, \mathbf{Z}) = \left\{ \begin{pmatrix} a & b \\ c & d \end{pmatrix} : a, b, c, d \in \mathbf{Z}, \, ad - bc = 1 \right\}. \tag{7.29}$$

This is the so called *duality group* of our theory, and acts on τ as

$$\tau \to \frac{a\tau + b}{c\tau + d}. \tag{7.30}$$

The transformations $\tau \to -1/\tau$ and $\tau \to \tau + 1$ are implemented by the matrices

$$S = \begin{pmatrix} 0 & 1 \\ -1 & 0 \end{pmatrix}, \quad T = \begin{pmatrix} 1 & 1 \\ 0 & 1 \end{pmatrix}. \tag{7.31}$$

In order to show that the effective theory (7.6) has this duality group, we have to analyse the first term in (7.6), which involves the chiral superfield A. Let us introduce $h(A) = \partial \mathcal{F}/\partial A$. In terms of this, $\tau(A) = \partial h(A)/\partial A$ and the scalar kinetic energy term becomes $\mathrm{Im} \int d^4\theta \, h(A)\overline{A}$. The dual theory corresponding to (7.28) is defined by a dual chiral field

$$A_D = h(A) = \frac{\partial \mathcal{F}}{\partial A}. \tag{7.32}$$

Under this transformation the scalar kinetic energy term transforms to

$$\mathrm{Im} \int d^4\theta \, h(A)\overline{A} = \mathrm{Im} \int d^4\theta \, h_D(A_D)\overline{A}_D, \tag{7.33}$$

where $h_D(h(A)) = -A$, and therefore retains its form. This transformation defines in particular a *dual prepotential* $\mathcal{F}_D(A_D)$ through the relation:

$$h_D(A_D) = \frac{\partial \mathcal{F}(A_D)}{\partial A_D} = -A, \tag{7.34}$$

and therefore

$$-\frac{1}{\tau(A)} = -\frac{1}{h'(A)} = h_D'(A_D) = \tau_D(A_D), \tag{7.35}$$

and the dual coupling constant is related to the original coupling constant τ through an S-duality transformation (7.28). In conclusion, given the effective

Lagrangian (7.6), which is written in an 'electric frame', (*i.e.*, in terms of the $U(1)$ photon of the underlying $SU(2)$ theory), one can write it through a duality transformation in a 'magnetic' frame, *i.e.*, in terms of a $U(1)$ gauge field which couples locally to magnetic monopoles. This transformation, induced by the matrix S, is usually called S-duality. In general we have an infinite number of frames or equivalent descriptions of the theory related by $\mathrm{Sl}(2, \mathbf{Z})$ transformations.

We can consider the effect of the full $\mathrm{Sl}(2, \mathbf{Z})$ group on A and $\mathcal{F}$, or equivalently on A and $A_D = \partial \mathcal{F}/\partial A$. The transformation (7.30) implies that

$$\begin{pmatrix} A_D \\ A \end{pmatrix} \rightarrow \begin{pmatrix} a & b \\ c & d \end{pmatrix} \begin{pmatrix} A_D \\ A \end{pmatrix}. \tag{7.36}$$

The transformation of $\mathcal{F}$ can be easily obtained from (7.36), or equivalently from

$$\begin{aligned} A'_D &= aA_D + bA \,, \\ A' &= cA_D + dA \,. \end{aligned} \tag{7.37}$$

The first equation can be integrated with respect to A' by using the second equation and the result is

$$\mathcal{F}' = \frac{1}{2} bd A^2 + \frac{1}{2} ac A_D^2 + bc A A_D + \mathcal{F} \,. \tag{7.38}$$

Finally, notice that the Kähler metric on moduli space,

$$ds^2 = \mathrm{Im}\, \frac{da_D}{du} \frac{d\bar{a}}{d\bar{u}}\, du d\bar{u} = -\frac{i}{2} \left(\frac{da_D}{du} \frac{d\bar{a}}{d\bar{u}} - \frac{d\bar{a}_D}{d\bar{u}} \frac{da}{du} \right) du d\bar{u} \tag{7.39}$$

is manifestly $\mathrm{Sl}(2, \mathbf{Z})$ invariant. Duality of the effective action implies that the correct expression for the central charge for an (n_e, n_m) dyon must be

$$Z = an_e + a_D n_m. \tag{7.40}$$

This can be easily seen as follows: by analysing the coupling of hypermultiplets of electric charge n_e to the chiral field A in the effective action

$$\sqrt{2} n_e \widetilde{M} A M, \tag{7.41}$$

one easily finds the expected mass $M = \sqrt{2}|a n_e|$. However, by $\mathrm{Sl}(2, \mathbf{Z})$ duality the mass of a monopole of charge n_m must be given by $M = \sqrt{2}|a_D n_m|$. This leads to (7.40) for an arbitrary dyon. Notice that in the semi-classical limit $a_D \sim \tau a^2$, and one recovers (7.21).

The Sl$(2, \mathbf{Z})$ covariance of the effective theory will be crucial for the structure of the Seiberg–Witten solution, and many of the quantities involved will be tensors under this transformation group. We will say that a quantity F has weight (n, m) under Sl$(2, \mathbf{Z})$ if under the transformation (7.30) it behaves as

$$F \to (c\tau + d)^n (c\bar{\tau} + d)^m F. \tag{7.42}$$

For example, u has weight $(0,0)$, since it is the vacuum expectation value of a gauge-invariant operator (and as it will become clear later, it is a good global coordinate on moduli space). Therefore it should not depend on the description we use. One can also use (7.38) to show that

$$\mathcal{F} - \frac{1}{2} a a_D \tag{7.43}$$

has weight $(0,0)$. Similarly, it is easy to see that du/da has weight $(-1,0)$.

7.3. Elliptic curves

The exact solution of Seiberg–Witten is given in terms of quantities associated with an elliptic curve. This curve, also called a *Seiberg–Witten curve*, plays a crucial role in the solution and its topological applications, and therefore it is convenient to stop at this point and give a brief summary of the properties of elliptic curves that will be needed in the following.

As is well known, every algebraic curve of genus one can be written in the Weierstrass form

$$Y^2 = 4X^3 - g_2 X - g_3 = 4(X - e_1)(X - e_2)(X - e_3), \tag{7.44}$$

where the coefficients g_2, g_3 are related to the roots e_i, $i = 1, 2, 3$, by the equations

$$g_2 = -4(e_2 e_3 + e_3 e_1 + e_1 e_2), \quad g_3 = 4 e_1 e_2 e_3. \tag{7.45}$$

The discriminant of the elliptic curve is defined by

$$\Delta = \prod_{i<j} (e_i - e_j)^2, \tag{7.46}$$

and in terms of g_2, g_3 it is given by

$$\Delta = \frac{1}{16}(g_2^3 - 27 g_3^2). \tag{7.47}$$

We can now use Abel's theorem, which states that an algebraic curve of genus one like (7.44) is of the form $\mathbf{C}/\Lambda$ for some lattice $\Lambda \subset \mathbf{C}$ with half periods ω_1, ω_3, such that $\mathrm{Im}(\omega_3/\omega_1) > 0$. The map from $\mathbf{C}/\Lambda$ to the curve (7.44) is given by

$$\psi(z) = (\wp(z), \wp'(z)) = (X, Y), \tag{7.48}$$

where we consider (X, Y) as inhomogeneous coordinates in $\mathbf{CP}^2$, and the Weierstrass function $\wp(z)$ verifies the differential equation

$$(\wp'(z))^2 = 4\wp(z)^3 - g_2\wp(z) - g_3. \tag{7.49}$$

Under this correspondence the half periods of the lattice Λ, ω_i, $i = 1, 2, 3$, $\omega_2 = \omega_1 + \omega_3$, are mapped to the roots $e_i = \wp(\omega_i)$ of the cubic equation in (7.44), and the differential dz on $\mathbf{C}/\Lambda$ is mapped to the abelian differential of the first kind dX/Y. The map (7.48) has an inverse given by

$$z = \psi^{-1}(p) = \int_\infty^p \frac{dX}{Y}, \tag{7.50}$$

which is defined modulo Λ. We can obtain an explicit expression for the inverse map (7.50) by doing the change of variable $t^2 = (e_1 - e_3)/(X - e_3)$, to obtain

$$z = -\frac{1}{\sqrt{e_1 - e_3}} F(\phi, k), \tag{7.51}$$

where $F(\phi, k)$ is the incomplete elliptic integral of the first kind, with modulus $k^2 = (e_2 - e_3)/(e_1 - e_3)$, and $\sin^2\phi = (e_1 - e_3)/(\wp(z) - e_3)$.

In fact, all these functions can be computed in terms of the roots e_i and elliptic functions. First of all we have the periods of the abelian differential dX/Y. We take the branch cut on the X-plane from e_3 to e_2, and from e_1 to infinity, so that the α_1 and α_2 cycles of the torus circle around $e_1 - e_2$ and around $e_3 - e_2$, respectively. Therefore the periods of the abelian differential dX/Y are given by

$$\begin{aligned}
2\omega_3 &= \oint_{\alpha_1} \frac{dX}{Y} = \int_{e_1}^{e_2} \frac{dX}{\sqrt{(X - e_1)(X - e_2)(X - e_3)}}, \\
2\omega_1 &= \oint_{\alpha_2} \frac{dX}{Y} = \int_{e_3}^{e_2} \frac{dX}{\sqrt{(X - e_1)(X - e_2)(X - e_3)}}.
\end{aligned} \tag{7.52}$$

Introducing now the complementary modulus $k'^2 = 1 - k^2$, we obtain a representation of the periods in terms of the complete elliptic integral of the first kind,

$$\omega_3 = \frac{i}{\sqrt{e_1 - e_3}} K'(k),$$

$$\omega_1 = \frac{1}{\sqrt{e_1 - e_3}} K(k), \tag{7.53}$$

where $K'(k) \equiv K(k')$. We will also need the Weierstrass ζ function which is defined by the equation $\zeta'(z) = -\wp(z)$. Because of this property we have that

$$\eta_i \equiv \zeta(\omega_i) = - \int_{\omega_j}^{\omega_2} dz \, \wp(z), \qquad i, j = 1, 3, \ j \neq i, \tag{7.54}$$

hence their values at the half periods can be computed in terms of complete elliptic integrals,

$$\eta_3 = - i \frac{e_3}{\sqrt{e_1 - e_3}} K'(k) - i\sqrt{e_1 - e_3} E'(k),$$

$$\eta_1 = - \frac{e_1}{\sqrt{e_1 - e_3}} K(k) + \sqrt{e_1 - e_3} E(k), \tag{7.55}$$

where $E'(k) \equiv E(k')$. These periods satisfy the so called Legendre's relation,

$$\eta_1 \omega_3 - \eta_3 \omega_1 = \frac{\pi i}{2}. \tag{7.56}$$

When the discriminant of an elliptic curve vanishes the curve becomes singular. It is easy to see that in such a situation at least one of the periods of the lattice goes to infinity, and geometrically this indicates that the corresponding cycle of the torus has 'pinched' and the curve has a node. Let us analyse the structure of the curve when $\omega_1 = \infty$ while ω_3 remains finite. It is easy to see that the roots are then given by

$$e_1 = e_2 = -\frac{e_3}{2} = -\frac{3g_3}{2g_2}. \tag{7.57}$$

Elliptic functions degenerate to trigonometric functions in this limit (see Appendix B), and by using that $e_3 = \wp(\omega_3)$, we find that the finite period is given by

$$\left(\frac{\pi}{2\omega_3}\right)^2 = \frac{9g_3}{2g_2}, \tag{7.58}$$

and evaluating $\zeta(\omega_3)$ leads to

$$2\eta_3 \omega_3 = \frac{\pi^2}{6}. \tag{7.59}$$

When *both* periods go to infinity one has that $e_i = g_2 = g_3 = 0$, and the curve has a cusp singularity $y^2 = 4x^3$. In that case elliptic functions further degenerate to rational functions.

7.4. The exact solution of Seiberg and Witten

The exact answer for the low energy effective action of pure $\mathcal{N} = 2$ supersymmetric Yang–Mills theory was given by Seiberg and Witten in terms of an auxiliary elliptic curve which will play a very important role in what follows. It is useful to rescale

$$a \to a/2, \quad \tau \to 2\tau \tag{7.60}$$

so that $\tau = \theta/\pi + (8\pi i/g^2)$. With these conventions the Seiberg–Witten curve is

$$y^2 = x^3 - ux^2 + \frac{1}{4}x, \tag{7.61}$$

in units where $\Lambda = 1$. This elliptic curve describes topologically a torus. To further specify the solution, we need another ingredient, which is an abelian differential λ_{SW} on the elliptic curve, with the property that

$$\frac{d\lambda_{\mathrm{SW}}}{du} = \frac{\sqrt{2}}{8\pi} \frac{dx}{y}. \tag{7.62}$$

This differential is usually called the *Seiberg–Witten differential*. The exact expression for the LEEA is completely determined by the expressions:

$$a_D = \oint_{\alpha_1} \lambda_{\mathrm{SW}}, \qquad a = \oint_{\alpha_2} \lambda_{\mathrm{SW}}. \tag{7.63}$$

This determines a, a_D in terms of u, and therefore determines implicitly the prepotential $\mathcal{F}(a)$ through $a_D = \partial \mathcal{F}/\partial a$.

It is not difficult to find an explicit expression for the Seiberg–Witten differential. By integrating (7.62) directly one finds:

$$\lambda_{\mathrm{SW}} = -\frac{\sqrt{2}}{4\pi} \frac{y}{x^2} dx, \tag{7.64}$$

which, up to an exact differential $\sqrt{2}/(2\pi)d(y/x)$, is given by

$$\lambda_{\mathrm{SW}} = \frac{\sqrt{2}}{8\pi} \frac{dx}{y} (2u - 4x). \tag{7.65}$$

Let us give a more explicit expression for the above integrals. The first thing to do is to write the Seiberg–Witten curve in the Weierstrass form (7.44). This curve (as well as the generalizations to theories with matter) has the form:

$$y^2 = x^3 + Bx^2 + Cx + D, \tag{7.66}$$

where the coefficients B, C, and D depend on the gauge-invariant parameter u and the dynamical scale of the theory Λ which we have set equal to one. To put this curve in the Weierstrass form it suffices to redefine the variables as

$$y = 4Y, \qquad x = 4X - \frac{1}{3}B, \tag{7.67}$$

and the curve (7.66) now has the form given in (7.44) with

$$g_2 = -\frac{1}{4}\left(C - \frac{1}{3}B^2\right), \qquad g_3 = -\frac{1}{16}\left(D + \frac{2B^3}{27} - \frac{CB}{3}\right). \tag{7.68}$$

Notice that with the redefinition given in (7.67), the abelian differential of the first kind is $dX/Y = dx/y$. In the case of the Seiberg–Witten curve (7.61) it is easy to find that

$$g_2 = \frac{1}{4}\left(\frac{u^2}{3} - \frac{1}{4}\right), \qquad g_3 = \frac{1}{48}\left(\frac{2u^3}{9} - \frac{u}{4}\right), \tag{7.69}$$

so that the discriminant of the curve is given by

$$\Delta = \frac{1}{4096}(u^2 - 1), \tag{7.70}$$

and the roots e_i have the explicit expression:

$$e_1 = \frac{u}{24} + \frac{\sqrt{u^2 - 1}}{8}, \qquad e_2 = \frac{u}{24} - \frac{\sqrt{u^2 - 1}}{8}, \qquad e_3 = -\frac{u}{12}. \tag{7.71}$$

Of course, at this point this labeling only holds up to permutation, but we will show in a moment that with the above conventions this is the right choice.

Notice that

$$\frac{da_D}{du} = \frac{\sqrt{2}}{4\pi}\omega_3, \qquad \frac{da}{du} = \frac{\sqrt{2}}{4\pi}\omega_1, \tag{7.72}$$

and therefore

$$\tau = \frac{da_D}{da} = \frac{\omega_3}{\omega_1}, \tag{7.73}$$

i.e., the effective gauge coupling constant is the τ parameter of the elliptic curve. In particular, $\operatorname{Im}\tau > 0$, as required for the positivity of the Kähler metric (7.8) on moduli space.

In order to compute a and a_D we have to compute the integrals:

$$a_D = \frac{\sqrt{2}}{\pi}\int_{\omega_1}^{\omega_2} dz\left(\frac{u}{6} - 4\wp(z)\right),$$
$$a = \frac{\sqrt{2}}{\pi}\int_{\omega_3}^{\omega_2} dz\left(\frac{u}{6} - 4\wp(z)\right). \tag{7.74}$$

Using (7.54) and $\zeta(\omega_2) = \zeta(\omega_1) + \zeta(\omega_3)$ we obtain

$$a_D = \frac{\sqrt{2}}{\pi}\left(4\zeta(\omega_3) + \frac{u}{6}\omega_3\right),$$
$$a = \frac{\sqrt{2}}{\pi}\left(4\zeta(\omega_1) + \frac{u}{6}\omega_1\right), \tag{7.75}$$

The use of elliptic functions leads to some useful results in Seiberg–Witten theory. For example, using Legendre's relation (7.56) one finds that

$$a\left(\frac{da_D}{du}\right) - a_D\left(\frac{da}{du}\right) = \frac{i}{\pi} \tag{7.76}$$

which after integration leads to

$$\mathcal{F} - \frac{1}{2}aa_D = -\frac{i}{2\pi}u, \tag{7.77}$$

which confirms that u has weight $(0,0)$ under $\mathrm{Sl}(2,\mathbf{Z})$.

Now we will analyse in more detail the exact solution of Seiberg and Witten by first checking that it reproduces the known weak coupling behavior, and then by looking at the strong coupling regime.

We first look at the weak coupling regime, the regime where $u \to \infty$. In this case one finds:

$$e_1 = \frac{u}{6} - \frac{1}{16u} + \cdots, \qquad e_2 = -\frac{u}{12} + \frac{1}{16u} + \cdots. \tag{7.78}$$

Now using the expansions of the elliptic integral of the first kind for small k^2,

$$K(k) = \frac{\pi}{2}\left\{1 + \frac{1}{4}k^2 + \cdots\right\},$$
$$K'(k) = \log\frac{4}{k} + \frac{1}{4}\left(\log\frac{4}{k} - 1\right)k^2 + \cdots, \tag{7.79}$$

we find

$$\omega_3 = \frac{2i}{\sqrt{u}}\left(\log u + 3\log 2 + \mathcal{O}\left(\frac{1}{u}\right)\right),$$
$$\omega_1 = \frac{\pi}{\sqrt{u}}\left(1 + \mathcal{O}\left(\frac{1}{u}\right)\right), \tag{7.80}$$

and

$$\tau(u) = \frac{2i}{\pi}\left(\log u + 3\log 2\right) + \mathcal{O}\left(\frac{1}{u}\right) \tag{7.81}$$

in agreement with the one-loop result. By using the expansion of the complete elliptic integral of the second kind,

$$E(k) = \frac{\pi}{2}\left\{1 - \frac{1}{4}k^2 + \cdots\right\},$$
$$E'(k) = 1 + \frac{1}{2}\left(\log\frac{4}{k} - \frac{1}{2}\right)k^2 + \cdots, \tag{7.82}$$

one can derive the leading terms of the expansions of $\zeta(\omega_{3,1})$,

$$\zeta(\omega_3) = \frac{i}{6}\sqrt{u}\log u - \frac{i}{6}\sqrt{u} + \cdots,$$
$$\zeta(\omega_1) = \frac{\pi}{12}\sqrt{u} + \cdots, \tag{7.83}$$

and use this to check the weak coupling expressions

$$a = \frac{\sqrt{2u}}{2}\left(1 + \mathcal{O}\left(\frac{1}{u}\right)\right),$$
$$a_D = \frac{i}{\pi}\sqrt{2u}\left(\log u + \mathcal{O}\left(\frac{1}{u}\right)\right). \tag{7.84}$$

Let us now analyse in more detail what happens at other points in the u-plane. There are clearly some special points in moduli space at which the curve (7.61) becomes singular. As we discussed at the end of section 3, this happens when two roots coincide and the discriminant vanishes. In the case of the Seiberg–Witten curve the singularities occur at $u = \pm 1$ in moduli space, where $e_1 = e_2$. Let us first consider the point $u = 1$. Since

$$k^2 = \frac{u - \sqrt{u^2 - 1}}{u + \sqrt{u^2 - 1}} \tag{7.85}$$

is one at $u = 1$, the period ω_1 diverges. Using (7.58) and (7.59) it is easy to see that at this point

$$a_D = 0. \tag{7.86}$$

After expanding a_D, a around $u = 1$, also known as the monopole point, as will become clear below:

$$a_D = \frac{i}{2}(u - 1) + \mathcal{O}(u - 1),$$
$$a = -\frac{1}{4\pi}(u - 1)\log(u - 1) + \mathcal{O}(u - 1), \tag{7.87}$$

one finds that the dual coupling can be expanded as

$$\tau_D = -\frac{i}{2\pi}\log a_D + \mathcal{O}(a_D). \tag{7.88}$$

Using the general expression for the β-function of a $U(1)$ theory coupled to Weyl fermions of charges Q_f and scalars with charge Q_s

$$\beta(g) = \frac{g^3}{16\pi^2}\left(\sum_f \frac{2}{3}Q_f^2 + \sum_s \frac{1}{6}Q_s^2\right), \tag{7.89}$$

we see that the effective coupling (7.88) behaves as the coupling of a $\mathcal{N} = 2$ supersymmetric $U(1)$ Yang–Mills theory coupled to a massive $\mathcal{N} = 2$ supersymmetric hypermultiplet of charge 1 and mass $\sim a_D$. Therefore what is happening is the following: the $\mathcal{N} = 2$ supersymmetric Yang–Mills theory has in its spectrum a magnetic monopole with charge $(n_m, n_e) = (1, 0)$ and mass $M = \sqrt{2}|a_D|$. In the semi-classical regime, when $u, a \gg 1$, the monopole is quite massive, and its mass behaves as $\sim |u|^{1/2} \log |u|$. However, as we go to strong coupling the monopole becomes lighter and precisely at $u = 1$ it becomes exactly *massless*. The fact that the the Seiberg–Witten solution has an apparent singularity at $u = 1$ has now a simple explanation: to obtain the LEEA one has to integrate out massive degrees of freedom. In particular, at generic points in the u-plane we have to integrate out as well the massive monopoles in the spectrum. However, integrating out a massless particle leads to singularities in the effective action (as we saw in the simple example of the $W^{\pm}$ bosons), and at $u = 1$ what we are seeing is precisely the singularity owing to integrating out the massless monopole. This is shown very explicitly in (7.88), where we see a logarithmic singularity in the one-loop effective coupling as the mass of the monopole running around the loop goes to zero.

A similar story takes place at $u = -1$: what becomes massless there is the dyon with quantum numbers $(n_e, n_m) = (1, 1)$. Although this is commonly referred to as a dyon (and we will do the same), it, is in fact, a monopole: owing to Witten's effect the effective electric charge is not n_e, but $q_e = n_e + \mathrm{Re}\,\tau n_m$. Using the Seiberg–Witten exact solution one can check that $q_e = 0$ at $u = -1$, so the particle becoming massless at $u = -1$ is physically a monopole, as it should be in view of the $u \to -u$ symmetry.

The picture that emerges from the Seiberg–Witten solution is then the following: the quantum moduli space of the $\mathcal{N} = 2$ supersymmetric theory is still the u-plane, but with a corrected Kähler metric given by a non-trivial prepotential, and with a singular behavior not only at infinity, but also at $u = \pm 1$. On this u-plane there is a (flat) $\mathrm{Sl}(2, \mathbf{Z})$ vector bundle, and (a_D, a) can be regarded as a section of this bundle. Different choices of the section are related by $\mathrm{Sl}(2, \mathbf{Z})$ transformations.

There is a subgroup of $\mathrm{Sl}(2, \mathbf{Z})$ that has a special significance, and it is called the *monodromy group*. This arises as follows: the u-plane has three punctures, at $u_* = \infty, 1, -1$. The punctures at $u = \pm 1$ correspond to the singular behavior owed to the massless monopole and massless dyon, as we have discussed, whilst at $u = \infty$ we have the semi-classical region governed by the one-loop beta function. Since we have a flat bundle the homotopy group of the u-plane (which is generated by the non-trivial one-cycles around the punctures) gives a subgroup of the structure group of the bundle $\text{Sl}(2, \mathbf{Z})$. This subgroup is the monodromy group, and it is also called the congruence subgroup of $\text{Sl}(2, \mathbf{Z})$ associated with the Seiberg–Witten elliptic curve. An explicit presentation of this group can be obtained as follows: at $u = \infty$ the effect of $u \to e^{2\pi i}u$ on a, a_D can be deduced from (7.84), and is given by the $\text{Sl}(2, \mathbf{Z})$ transformation

$$M_\infty = \begin{pmatrix} -1 & 4 \\ 0 & -1 \end{pmatrix}. \tag{7.90}$$

On the other hand, near $u = 1$ the effect of $u - 1 \to e^{2\pi i}(u - 1)$ can be read from (7.87), and one finds

$$M_1 = \begin{pmatrix} 1 & 0 \\ -1 & 1 \end{pmatrix}. \tag{7.91}$$

The monodromy M_{-1} at $u = -1$ can be obtained from the requirement that $M_1 M_{-1} = M_\infty$.

The matrices (7.90) and (7.91) generate the congruence subgroup $\Gamma^0(4)$ of $\text{Sl}(2, \mathbf{Z})$, which is defined by

$$\Gamma^0(4) = \{\begin{pmatrix} a & b \\ c & d \end{pmatrix} \in \text{Sl}(2, \mathbf{Z}) : b \equiv 0 \,\text{mod}\, 4\}. \tag{7.92}$$

This is the congruence subgroup of the Seiberg–Witten curve (7.61).

Since the Seiberg–Witten solution tells us the exact prepotential we can write the LEEA at any point in moduli space. At a generic point of the u-plane the low energy degrees of freedom are the fields of an $\mathcal{N} = 2$ supersymmetric vector multiplet, and the LEEA is given by (7.6). However, near $u = 1, -1$ this action becomes singular because we are integrating out a multiplet which is becoming massless. To cure the singularity we have to add this multiplet to the effective action. Suppose, for example, that we are near $u = 1$. The monopole hypermultiplet couples locally to the dual $\mathcal{N} = 2$ supersymmetric vector multiplet $\mathcal{A}_D$. Therefore, the LEEA will be the action of a $U(1)$ $\mathcal{N} = 2$ supersymmetric theory coupled to an $\mathcal{N} = 2$ supersymmetric hypermultiplet with charge 1. The Lagrangian is then (7.6) written in terms of the $\mathcal{A}_D$, together with

$$\int d^2\theta d^2\bar{\theta} \left(M^\dagger e^{2V_D} M + \widetilde{M}^\dagger e^{-2V_D} \widetilde{M}\right) + \sqrt{2}\left(\int d^2\theta \widetilde{M} \mathcal{A}_D M + \text{h.c.}\right). \tag{7.93}$$

There is, however, an important subtlety: the LEEA in magnetic variables is determined by the dual prepotential $\mathcal{F}_D(a_D)$. This prepotential is obtained after integrating out the monopole, and this is what leads to a logarithmic divergence in the τ_D coupling near $u = 1$. If we do *not* integrate out the monopole, but we include it in the action, the prepotential should be modified (and in particular the singularity of the modified prepotential should be smoothed out). The corrected prepotential $\widetilde{\mathcal{F}}_D(a_D)$ turns out to be determined by

$$\frac{d^2}{da_D^2}\widetilde{\mathcal{F}}_D(a_D) = \tau_D - \frac{1}{2\pi i}\log a_D, \tag{7.94}$$

and it is smooth. We will justify this in Chapter 9.

7.5. The Seiberg–Witten solution in terms of modular forms

In this section we provide some extra details about the Seiberg–Witten curve which will be important in the analysis of the u-plane integral.

So far we have parametrized the quantum moduli space by the u-variable. Since this moduli space describes a family of elliptic curves, we can parametrize it in terms of the τ modulus of the tori. Of course, there is an infinite family of frames which we can choose. In this section τ will denote the electric frame, which is the most convenient one for the semi-classical region. Since $\mathrm{Im}\,\tau > 0$, τ lives in the upper half-plane of the complex plane $\mathcal{H}$. However, two values of τ which are related by an element of the monodromy group of the curve should be regarded as equivalent. In order to find the moduli space of the elliptic curve, we then have to take the quotient of $\mathcal{H}$ by the monodromy group. If the monodromy group was $\mathrm{Sl}(2, \mathbf{Z})$ we could take the standard fundamental domain

$$\mathcal{F} = \left\{\tau \in \mathcal{H} : -\frac{1}{2} \leq \mathrm{Re}\,\tau \leq \frac{1}{2}, |\tau| \geq 1\right\} \tag{7.95}$$

as our moduli space. However, the monodromy group of an elliptic curve is usually a subgroup of $\mathrm{Sl}(2, \mathbf{Z})$. In the case of the Seiberg–Witten curve we have seen that the monodromy group is $\Gamma^0(4)$. This is a congruence subgroup of $\mathrm{Sl}(2, \mathbf{Z})$ of index 6, and $\mathrm{Sl}(2, \mathbf{Z})$ can be written as a union of cosets:

$$\mathrm{Sl}(2, \mathbf{Z}) = \bigcup_{i=1}^{6} \alpha_i^{-1}\Gamma^0(4), \tag{7.96}$$

where

$$\alpha_1 = 1, \quad \alpha_2 = T, \, \alpha_3 = T^2,$$
$$\alpha_4 = T^3, \, \alpha_5 = S, \, \alpha_6 = T^2 S. \tag{7.97}$$

It is not difficult to prove that $\cup_{i=1}^{6} \alpha_i \mathcal{F}$ is a fundamental domain for $\Gamma^0(4)$. We then find that the moduli space of the Seiberg–Witten curve (when parametrized by τ) is given by the fundamental domain of $\Gamma^0(4)$:

$$\Gamma^0(4) \backslash \mathcal{H} = \mathcal{F} \cup T \cdot \mathcal{F} \cup T^2 \cdot \mathcal{F} \cup T^3 \cdot \mathcal{F} \cup S \cdot \mathcal{F} \cup T^2 S \cdot \mathcal{F}. \tag{7.98}$$

The first four domains give the region of the cusp at $\tau \to i\infty$ and correspond to the semi-classical region. The region $S \cdot \mathcal{F}$ surrounds the cusp near $\tau = 0$ and will be referred to as the monopole cusp (notice that $\tau = 0$ is the value of the electric coupling constant at $u = 1$). The region $T^2 S \cdot \mathcal{F}$ surrounds the cusp near $\tau = 2$ and corresponds to the massless dyon at $u = -1$.

One can, in fact, write all the quantities involved in the Seiberg–Witten solution in terms of (generalized) modular forms with respect to the congruence subgroup $\Gamma^0(4)$. Notice that the action of the modular group that we are considering here is conceptually different from the action of $\mathrm{Sl}(2, \mathbf{Z})$ which we have discussed so far. The $\mathrm{Sl}(2, \mathbf{Z})$ action discussed in section 2 is a change of frame for each point in the moduli space, whilst the $\Gamma^0(4)$ transformations in (7.98) relate different regions of the moduli space.

In order to write the relevant quantities explicitly, the starting point is the relation between the roots of the Weierstrass cubic (7.44) and the theta functions:

$$e_1 = \left(\frac{\pi}{2\omega_1}\right)^2 \frac{1}{3}(\vartheta_3^4(\tau) + \vartheta_4^4(\tau)),$$

$$e_2 = \left(\frac{\pi}{2\omega_1}\right)^2 \frac{1}{3}(\vartheta_2^4(\tau) - \vartheta_4^4(\tau)), \tag{7.99}$$

$$e_3 = -\left(\frac{\pi}{2\omega_1}\right)^2 \frac{1}{3}(\vartheta_2^4(\tau) + \vartheta_3^4(\tau)).$$

Our conventions on Jacobi theta functions can be found in Appendix B. It follows from (7.99) that the discriminant of the elliptic curve is given by

$$\Delta = \left(\frac{2\pi}{\omega_1}\right)^{12} \eta^{24}(q), \tag{7.100}$$

where we have used the property $\vartheta_2^4(\tau)\vartheta_3^4(\tau)\vartheta_4^4(\tau) = 2^4 \eta^{12}(q)$. By using the explicit expressions for the periods and τ in the electric frame, at large u, one can check that the identification between the roots and the theta functions given in (7.99) is the correct one.

The next step is to find an expression for the period ω_1 and for u in terms of modular forms. This is easily done by using that $e_3 = -u/12$, and that

$$e_2 - e_1 = \frac{1}{4}\sqrt{u^2 - 1} = \left(\frac{\pi}{2\omega_1}\right)^2 \vartheta_4^4(\tau). \tag{7.101}$$

Using these results one can obtain:

$$u = \frac{1}{2}\frac{\vartheta_2^4(\tau) + \vartheta_3^4(\tau)}{(\vartheta_2\vartheta_3)^2} \tag{7.102}$$

and

$$h(\tau) = \frac{da}{du} = \frac{1}{2}\vartheta_2\vartheta_3. \tag{7.103}$$

In order to find expressions in terms of modular forms for the remaining quantities involved in the Seiberg–Witten solution, the only extra ingredient we need is the relation between the value of the ζ function at the half period and the Eisenstein series,

$$\zeta(\omega_1) = \frac{\pi^2}{12\omega_1}E_2(\tau), \tag{7.104}$$

which gives

$$a = \frac{1}{6}\left(\frac{2E_2 + \vartheta_2^4 + \vartheta_3^4}{\vartheta_2\vartheta_3}\right). \tag{7.105}$$

By performing an S-duality transformation, one can easily derive the expressions for the dual quantities listed in Appendix B (except for a_D, τ_D and q_D, these quantities are denoted by the subindex M, which refers to 'monopole'.)

Bibliographical notes

- The Seiberg–Witten solution of $SU(2)$ $\mathcal{N} = 2$ supersymmetric gauge theories in four dimensions is presented in the classical papers [23] (for supersymmetric Yang–Mills theory) and in [24] (for the theories with flavors, $i.e.$, supersymmetric quantum cromodynamics). These results have been reviewed in many places, amongst them [42][82][83][84].

- The results of Seiberg and Witten on $\mathcal{N} = 2$ supersymmetry build on previous results by Seiberg on theories with $\mathcal{N} = 1$ supersymmetry, see, for example, [85]. The first analysis of the effective prepotential of $\mathcal{N} = 2$ supersymmetric Yang–Mills theory was performed in [86], which discusses the one-loop result and the instanton corrections to $\mathcal{F}(a)$. The computation of instanton corrections has become a crucial aspect of the non-perturbative physics since the seminal work of 't Hooft [87]. Since the exact solution of Seiberg and Witten predicts the values of all the instanton coefficients in (7.19), there have been many works where these predictions have been checked by direct computation, see [88] for a review. An analytic computation of all these coefficients was achieved in [89], and this, together with some extra work in [90], leads to a derivation of the Seiberg and Witten solution. For details on $U(1)$ anomalies in the functional integral, see, for example, [91].

- There have been several hundreds of papers analysing different aspects of the Seiberg–Witten solution, and we cannot do justice to all of them. The identity (7.77) was first obtained in [92]. A discussion of the dyon point in the light of Witten's effect [93] is presented in [94].

- A standard reference in the theory of elliptic functions is [95]. Elliptic curves and modular forms are discussed, for example, in [96], and we list some useful formulae in Appendix B. For the use of elliptic functions in the Seiberg–Witten solution we follow mainly [97] (see also [98] and [99]). Expressions in terms of modular forms are listed in [61].

Chapter 8

The u-plane Integral

In this chapter we put together all the previous ingredients in order to compute the Donaldson–Witten generating function by using the Seiberg–Witten solution for the LEEA. This computation involves in general an integration over the moduli space of vacua parametrized by u. This u-plane integral was analysed in detail by Moore and Witten. Although this integral is different from zero only when $b_2^+ = 1$, it turns out that it gives a very effective method of deriving the expression for the Donaldson invariants in the general case.

8.1. The basic principle (or, 'Coulomb + Higgs=Donaldson')

Now that we have a low energy description of $\mathcal{N} = 2$ supersymmetric Yang–Mills theory, we can use it to compute the functional integral of Donaldson–Witten theory, and therefore to compute the Donaldson invariants. In the physical theory, using the effective theory in the two-derivative approximation only gives approximate results, which are valid for energies much lower than the dynamical scale Λ. For example, the two-derivative approximation to the pion Lagrangian of quantum cromodynamics gives only the bare bones of the pion–pion scattering amplitude. But in a topological theory the correlation functions of observables do not depend on the scale, and therefore the low energy approximation should be exact, without any further correction. We conclude that the two-derivative LEEA, after twisting, is all we need in order to compute Z_{DW} in (5.48). This can be stated in yet another way: since Donaldson–Witten theory is topological, we can compute correlation functions either at high energy (small distances), or at low energy (long distances). The underlying physical theory, as we have seen in the last chapter, is asymptotically free, therefore very high energy means very small coupling constant, and we know from the analysis in Chapter 5 that the semiclassical computation of correlation functions leads directly to the definition of Donaldson invariants. On the other hand, at very low energy we can use the LEEA, and this leads to a twisted abelian theory with monopoles and to the Seiberg–Witten invariants. It follows that Donaldson invariants can be computed by using a $U(1)$ twisted $\mathcal{N} = 2$ supersymmetric theory, possibly coupled to monopole matter fields.

What is the general structure of the computations in the low energy theory? In a physical theory in Minkowski space, one does not integrate over zero-modes which are not normalizable. These modes are rather parameters that specify our theory. For example, the u-plane description which we have presented in the previous section gives, in fact, a family of theories parametrized by u, since in a non-compact space the zero-mode of the scalar field u is not normalizable. However, on a compact manifold a scalar zero-mode *is* normalizable and we have to integrate over it. Indeed, computing correlation functions on a compact manifold requires one to perform a sum over the contributions of all the vacua. This means that the evaluation of Z_{DW} in the effective theory involves an integration over the u-plane, *i.e.*, over the Coulomb branch of $\mathcal{N} = 2$ supersymmetric Yang–Mills theory.

However, there are two points in the u-plane where something special happens: these are the points $u = \pm 1$ at which there are extra degrees of freedom becoming massless. At these points the contribution from the effective theory has to be different, since for example we will have to perform a functional integral over these extra degrees of freedom. We then expect that Z_{DW} looks like

$$Z_{\mathrm{DW}} = Z_u + Z_{u=1} + Z_{u=-1}, \tag{8.1}$$

where the first summand is the integral over the u-plane, and the second and third terms come from the localized contributions of the monopole and the dyon singularities. This basic structure can be summarized by the principle 'Donaldson = Coulomb + Higgs'. 'Higgs' refers to the scalars in the monopole and dyon multiplet, of course. The rest of this chapter is devoted to explaining (8.1) in detail.

8.2. Effective topological quantum field theory on the u-plane

It turns out that it is more convenient to start our analysis with Z_u. The reason is that (modulo some subtleties which we will explain in due time) the computation of Z_u involves just the twisted version of the effective Lagrangian (7.6). This is done as in Donaldson–Witten theory, with the simplification that the theory described by (7.6) is abelian, but with the complication that it has an arbitrary prepotential (the abelian version of Donaldson–Witten theory corresponds to the prepotential $\mathcal{F} = a^2/2$). In any case, the $\overline{Q}$ symmetry is precisely the abelian version of (5.29) (after the rescaling (7.60)). We can easily obtain from (7.10) the Lagrangian density:

$$
\begin{aligned}
\mathcal{L} =\, & \frac{i}{16\pi}\left(\overline{\tau}F_+ \wedge F_+ + \tau F_- \wedge F_-\right) + \frac{1}{8\pi}\mathrm{Im}\tau da \wedge *d\overline{a} - \frac{1}{8\pi}(\mathrm{Im}\tau)D \wedge *D \\
& - \frac{1}{16\pi}\tau\psi \wedge *d\eta + \frac{1}{16\pi}\overline{\tau}\eta \wedge d*\psi + \frac{1}{8\pi}\tau\psi \wedge d\chi - \frac{1}{8\pi}\overline{\tau}\chi \wedge d\psi \\
& + \frac{i\sqrt{2}}{16\pi}\frac{d\overline{\tau}}{da}\eta\chi \wedge (D_+ + F_+) - \frac{i\sqrt{2}}{2^7\pi}\frac{d\tau}{da}(\psi \wedge \psi) \wedge (F_- + D_+) \\
& + \frac{i}{3 \cdot 2^{11}\pi}\frac{d^2\tau}{da^2}\psi \wedge \psi \wedge \psi \wedge \psi - \frac{\sqrt{2}i}{3 \cdot 2^5\pi}\{\overline{Q}, \frac{d\overline{\tau}}{da}\chi_{\mu\nu}\chi^{\nu\lambda}\chi_\lambda{}^\mu\}\sqrt{g}d^4x\,.
\end{aligned}
\tag{8.2}
$$

This can be also written in terms of the fourth descendant of the prepotential:

$$
\begin{aligned}
\mathcal{L} =\, & \frac{i}{6\pi}G^4\mathcal{F}(a) + \frac{1}{16\pi}\{\overline{Q}, \overline{\mathcal{F}}''\chi(D + F_+)\} - \frac{i\sqrt{2}}{32\pi}\{\overline{Q}, \overline{\mathcal{F}}'d*\psi\} \\
& - \frac{\sqrt{2}i}{3 \cdot 2^5\pi}\{\overline{Q}, \overline{\mathcal{F}}'''\chi_{\mu\nu}\chi^{\nu\lambda}\chi_\lambda{}^\mu\}\sqrt{g}d^4x,
\end{aligned}
\tag{8.3}
$$

where G is the operator defined in (5.32).

However, the effective action of the topological theory has some extra terms which can not be derived from the physical theory (7.10) in Minkowski space. It is well known that effective field theories in curved space contain extra couplings to the background metric, and these are the extra terms which we are missing in (8.2). When the theory is topological the extension of the theory to curved space is restricted by $\overline{Q}$-invariance. The extra terms are forced to have the structure

$$
\int d^4x\sqrt{g}\, g(u)F(g_{\mu\nu}),
\tag{8.4}
$$

where $F(g_{\mu\nu})$ is some functional of the metric, and $\overline{Q}$-invariance implies that $g(u)$ should be a holomorphic function of u (since $[\overline{Q}, \overline{a}] \neq 0$). Moreover, since the theory is topological the functionals of the metric have to give rise to topological invariants of the four-manifold. The only topological invariants that can be written as local functionals of the metric are

$$
\int_X \mathrm{Tr}(R \wedge *R), \qquad \int_X \mathrm{Tr}(R \wedge R),
\tag{8.5}
$$

which are proportional to the Euler characteristic χ and the signature σ, respectively. Therefore the integrand of the functional integral $\exp(-S_{\mathrm{eff}})$ will contain an extra factor

$$A(u)^{\chi} B(u)^{\sigma} \tag{8.6}$$

which can be regarded as a contribution to the effective measure of the low energy theory. Fortunately, the couplings in (8.6) can be determined by holomorphy and anomaly considerations. The basic idea is the following: as we have seen in the previous chapter, the microscopic theory has an anomalous $U(1)_R$ symmetry. This means that the quantum measure has a certain weight under a $U(1)_R$ transformation, which on a general four-manifold will depend on χ and σ. On the other hand, the effective measure of the low energy theory is its canonical measure times the factor (8.6). Therefore, requiring that the effective measure reproduces the anomaly of the microscopic measure may fix the value of (8.6).

Let us see how this works in some detail. Consider the theory at large u. The effective theory is a $U(1)$ gauge theory. Under the anomalous $U(1)_R$ symmetry the measure of the microscopic theory transforms as in (7.17). However, the Dirac operator has changed after twisting the theory (the spinors are now differential forms). The resulting operator is, in fact, the operator associated to the complex (2.46), and its index is given on a four-manifold and for the gauge group $SU(2)$ by:

$$-8k + \frac{3}{2}(\chi + \sigma). \tag{8.7}$$

In flat space the anomaly is just $8k$, and this is what explained the spontaneous breaking of the $U(1)_R$ symmetry. The gravitational part of this anomaly is $\frac{3}{2}(\chi + \sigma)$: the $U(1)$ photon multiplet and the $W^{\pm}$ contribute $\frac{1}{2}(\chi + \sigma)$ each. However, the canonical measure of the effective theory only gives $\frac{1}{2}(\chi + \sigma)$ (since it only contains the $U(1)$ photon multiplet), and the remaining anomaly should be reproduced by the effective interaction (8.6) (since it comes from integrating out the massive $W^{\pm}$ multiplets which carry the remaining anomaly). As u has R-charge 4 we must have (for $u \to \infty$):

$$A(u)^{\chi} B(u)^{\sigma} \sim u^{(\chi+\sigma)/4}. \tag{8.8}$$

Let us now consider the effective theory near $u = 1, -1$. Near $u = 1$ there is an extra light degree of freedom, the monopole. The measure of the twisted effective theory including the monopole now has R-charge

$$\frac{1}{2}(\chi + \sigma) - \frac{c_1(L)^2}{4} + \frac{\sigma}{4}. \tag{8.9}$$

Since the LEEA does not include the monopole, it misses the $\sigma/4$. But the full u-plane theory near $u = 1$ is obtained by integrating out the monopole, therefore the effective measure (8.6) has to reproduce this anomaly. Remember that in the monopole theory a_D has R-charge 2, and since $a_D \sim u - 1$ we have for $u \to 1$

$$A(u)^\chi B(u)^\sigma \sim (u - 1)^{\sigma/8}. \tag{8.10}$$

In the same way we have that

$$A(u)^\chi B(u)^\sigma \sim (u + 1)^{\sigma/8}, \tag{8.11}$$

for $u \to -1$. Looking at (8.8) and (8.11), we see that the function

$$B(u)^\sigma = \beta^\sigma (u^2 - 1)^{\sigma/8}, \tag{8.12}$$

where β is a constant, satisfies our requirements. Notice that this holomorphic function is invariant under duality transformations (since u has weight $(0,0)$), and it involves in a natural way the discriminant of the Seiberg–Witten curve.

The function $A(u)$ is trickier. It should not have zeros or poles at $u = \pm 1$, since the anomaly at those points does not have any χ-dependence. It turns out that

$$\left(\frac{du}{da}\right)^{\chi/2} \tag{8.13}$$

satisfies all the requirements. Notice that this is not invariant under duality transformations: it transforms with weight $-\chi/2$. When we choose the appropriate local coordinate, it is clear that du/da has not zeroes or poles at $u = \pm 1$ (for example, at the monopole point $du/da_D = -2i$). Since $u \sim a^2$ as $u \to \infty$, (8.13) behaves as $u^{\chi/4}$ in the semi-classical region. We then have

$$A(u)^\chi = \alpha^\chi \left(\frac{du}{da}\right)^{\chi/2}, \tag{8.14}$$

where α is a constant. We will see later that, in fact, the weight of $A(u)^\chi$ under $\mathrm{Sl}(2, \mathbf{Z})$ is the required one to have a consistent u-plane integral.

There is another topological term that we have to add to the effective action. This is the coupling

$$\frac{i}{4} \int_X F \wedge w_2(X), \tag{8.15}$$

where $w_2(X)$, the second Stiefel–Whitney class of X, is interpreted as a gravitational background. We will see that, in fact, such a term is required for
consistency of the u-plane integral (and this fixes the coefficient in (8.15)),
and is responsible for the property of the line bundle of the magnetic theory
giving rise to a Spin_c structure. It can be also derived directly by integrating
out the massive fermions in the twisted theory.

The main result of this discussion is that the effective measure of the low
energy field theory is given by the canonical measure multiplied by

$$A^\chi B^\sigma = \alpha^\chi \beta^\sigma (u^2 - 1)^{\sigma/8} \left(\frac{du}{da}\right)^{\chi/2}. \tag{8.16}$$

The values of α, β can be obtained by comparison with mathematical results.
It turns out that

$$\alpha^4 = -\frac{2}{\pi}, \quad \beta^4 = -\frac{16}{\pi}. \tag{8.17}$$

In addition to the measure and the action itself we have to include the
observables of the theory. In principle, the low energy operators are obtained
from the high energy operators by integrating out massive modes, and *a priori* it is not simple to obtain their explicit expressions. However, in the case
at hand we can use the topological structure to find the operators in the
effective theory. As we explained in detail in Chapter 5, the observables of
Donaldson theory (in the $SU(2)$ case) are the operator $\mathcal{O} = \mathrm{Tr}\,(\phi^2)$ together
with its topological descendants. In order to obtain the corresponding observables in the low energy theory we first notice that $\mathcal{O}$ corresponds (up to
a normalization) to u, as follows from (7.3). In fact, we will have

$$\mathcal{O} \to 2u, \tag{8.18}$$

where the normalization factor 2 is chosen to agree with the usual normalization in Donaldson theory. We now note that the effective theory is also
topological, and in particular it contains a descent operator G. Since the
operators in Donaldson theory are obtained by topological descent from $\mathcal{O}$,
the corresponding operators in the low energy theory are obtained from u
by using the canonical solution to the topological descent equations, *i.e.*, by

acting with the operator G on u. In this way we find

$$I_1(\gamma) \to \tilde{I}_1(\gamma) = a_1 \int_\gamma Gu = \frac{a_1}{4\sqrt{2}} \frac{du}{da} \int_\gamma \psi,$$

$$I_2(S) \to \tilde{I}_2(S) = \frac{i}{\pi\sqrt{2}} \int_S G^2 u \tag{8.19}$$

$$= \frac{i}{\pi\sqrt{2}} \int_S \left\{ \frac{1}{32} \frac{d^2 u}{da^2} \psi \wedge \psi - \frac{\sqrt{2}}{4} \frac{du}{da} (F_- + D_+) \right\}.$$

In this equation we have rescaled the action of G in the way prescribed by
(7.60). The overall normalizations of the observables are chosen again in order
to agree with the normalizations in Donaldson theory, and

$$a_1 = \pi^{-1/2} 2^{3/4} e^{-i\pi/4}. \tag{8.20}$$

In writing the above correspondence between high energy and low energy operators we have not taken into account an important subtlety: in general,
integrating out high-energy modes and taking products of operators are operations which do not commute. Therefore it is in general not true that

$$I_2(S_1) I_2(S_2) \longrightarrow \tilde{I}_2(S_1) \tilde{I}_2(S_2). \tag{8.21}$$

What happens is that the intersection points of S_1 and S_2 give singularities
in the propagators that can induce extra contributions, and in the low energy
theory we expect contact terms located at the intersection locus of the two-
dimensional homology classes. Instead of (8.21) we will, rather, have:

$$I_2(S_1) I_2(S_2) \to \tilde{I}_2(S_1) \tilde{I}_2(S_2) + \sum_{P \in S_1 \cap S_2} \epsilon_P T(P). \tag{8.22}$$

Here T is some operator that one should determine based on a series of re-
quirements. For example, (8.22) must be $\overline{Q}$-invariant, and since $\tilde{I}_2(S_1)\tilde{I}_2(S_2)$
is already $\overline{Q}$-invariant the operator T must be separately $\overline{Q}$-invariant. This
means that T is a holomorphic function of u, as are the contributions to the
effective measure which we determined above. We will see later that T must
have a very precise behavior under $\mathrm{Sl}(2, \mathbf{Z})$ transformations, and using this we
will be able to find it explicitly. We point out that the contact terms can not
be obtained directly from the Seiberg–Witten solution, and extra arguments
are needed to determine them. The main conclusion of this short discussion

is then that there is potentially in the effective theory a contact term with the form

$$\exp\big(T(u)S^2\big), \tag{8.23}$$

where we have written the geometric intersection in (8.22) as $(S,S) = S^2$. Since one-cycles do not generically intersect in a four-manifold, there will not be contact terms for the operators $I_1(\gamma)$.

8.3. Zero modes

At this point we have determined almost all the ingredients of the u-plane integral: the twisted action, the observables (together with their contact terms) and the effective measure. The u-plane integral contribution to the Donaldson–Witten generating functional has the form

$$Z_u = 2 \int [da\, d\bar{a}\, d\eta\, d\chi\, d\psi\, dD] A^\chi B^\sigma y^{-1/2} \exp\left[\frac{1}{8\pi}\int (\mathrm{Im}\tau) D \wedge *D\right]$$

$$\times \exp\left[-\frac{i\pi}{16\pi}\bar{\tau}F_+^2 - \frac{i\pi}{16\pi}\tau F_-^2 + \frac{\pi i}{4}(F, w_2(X))\right]$$

$$\times \exp\left[-\frac{i\sqrt{2}}{16\pi}\int \frac{d\bar{\tau}}{d\bar{a}}\eta \wedge (D_+ + F_+) + \frac{i\sqrt{2}}{2^7\pi}\int \frac{d\tau}{da}(\psi \wedge \psi) \wedge (F_- + D_+)\right.$$

$$+ \frac{1}{3 \cdot 2^{11}\pi i}\int \frac{d^2\tau}{da^2}\psi \wedge \psi \wedge \psi \wedge \psi + 2pu + S^2 T(u)$$

$$+ \left.\frac{i}{\sqrt{2\pi}}\int_S G^2 u + a_1 \int_\gamma Gu + \cdots\right]. \tag{8.24}$$

In the above expression we have skipped some terms (like the kinetic terms for the dynamic fields). The global factor of 2 is to correct a standard discrepancy between physical and mathematical computations of the invariants, since physicists divide by the order of the center of the gauge group in the Fadeev-Popov gauge fixing, whilst mathematicians do not.

In order to perform the above functional integral in the low energy theory, we have to divide the fields into zero-modes and quantum fluctuations:

$$a = a_0 + a',$$
$$\psi = \psi_0 + \psi',$$
$$\eta = \eta_0 + \eta',$$
$$\chi = \chi_0 + \chi',$$
$$A = A_0 + A', \tag{8.25}$$

and then we have to determine the nature and the measure of the zero-modes. The geometric content of the zero-modes is easy to obtain by looking at the

kinetic terms of the effective theory: a_0 is just a constant, A_0 is a $U(1)$
connection with harmonic curvature, η_0 is a constant Grassmann variable, ψ_0
is given by a Grassmann variable multiplied by a harmonic one-form, and χ_0
is a Grassmann variable multiplied by a harmonic self-dual two-form.

One would be tempted to follow the analysis we performed in the formal
study of topological quantum field theories and localize the functional integral
on supersymmetric configurations. However, in this case this is not a good
idea: the above theory is an effective theory with a $\mathrm{Sl}(2, \mathbf{Z})$ covariance, and
by localizing on classical solutions we break this covariance. However, it
was shown by Moore and Witten that, for $b_2^+ = 1$, *the only contributions
to the functional integral (8.24) come from the zero-modes.* This analysis is
based on the topological invariance of the theory under global rescalings of
the metric, $g \to t^2 g$ (notice that for $b_2^+ = 1$ Donaldson invariants depend
on the conformal class of the metric, and, in particular, they are invariant
under global scalings). The same analysis shows that for $b_2^+ > 1$ there are too
many zero-modes of the χ field to be soaked up, and the functional integral
vanishes. Therefore

$$Z_u = 0 \quad \text{for } b_2^+ > 1. \tag{8.26}$$

This does not mean that Donaldson–Witten theory is trivial for $b_2^+ > 1$: remember that Z_u is just the contribution to the final answer coming from the
'Coulomb' vacua, and there are extra pieces at the singular points $u = \pm 1$.
This vanishing result means that for manifolds of $b_2^+ > 1$ the Donaldson invariants are given entirely in terms of the 'Higgs' contribution $Z_{u=\pm 1}$. Another
outcome of the analysis of Moore and Witten is that for $b_2^+ = 0$ the functional
integral has one-loop contributions also. The analysis of the u-plane integral
for $b_2^+ = 0$ remains an interesting open problem.

Since the functional integral can be computed by zero-modes, we are
dealing effectively with a finite-dimensional integral. We just have to find the
appropriate measure for the fields and perform the integral. Owing to (8.26)
we can restrict ourselves to manifolds with $b_2^+ = 1$. On these manifolds there
is only one harmonic self-dual two-form ω (whose normalization is fixed by
$\omega^2 = 1$), and we will write

$$\chi = \chi_0 \omega + \chi', \qquad \psi = \sum_{i=1}^{b_1} c_i \beta_i + \psi', \tag{8.27}$$

where the $\{\beta_i\}_{i=1,\ldots,b_1}$ is a basis for the harmonic one-forms. Notice that
the kinetic terms for the fields in (8.2) are not canonical, since they involve

a factor $\operatorname{Im} \tau$. Therefore, the measure includes a factor $(\operatorname{Im} \tau)^{1/2}$ for the commuting fields and $(\operatorname{Im} \tau)^{-1/2}$ for the anticommuting fields (this is just the Jacobian for the change of variables that takes us to canonical kinetic terms). After combining all the factors we find that the measure for the zero-modes of the fields (except for the gauge field) is given by:

$$(\operatorname{Im} \tau)^{-b_1/2} da\, d\bar{a}\, d\eta_0\, d\chi_0 \prod_{i=1}^{b_1} dc_i. \tag{8.28}$$

It remains to discuss the zero-modes for the gauge field. This is a little bit trickier. The solutions of the equations of motion, $dF = d * F = 0$, are harmonic two-forms. The space of collective coordinates for the zero-modes will split into different topological sectors, corresponding to different line bundles T over the four-manifold X, with $c_1(T) = F/2\pi \in H^2(X, \mathbf{Z})$. Now we have to remember that the gauge bundle we are considering is an $SO(3)$ bundle V with a non-trivial Stiefel–Whitney class $w_2(V)$. On the u-plane the non-abelian gauge symmetry is broken down to $U(1)$, and this means that the $SO(3)$ bundle V has the structure (2.34). Since by (2.37) $c_1(T)$ is congruent to $w_2(V)$ mod 2 we will write

$$F = 4\pi\lambda, \tag{8.29}$$

where $\lambda \in H^2(X, \mathbf{Z}) + \frac{1}{2}w_2(V)$. We denote by Γ the lattice of integral two-forms $H^2(X, \mathbf{Z})$.

For each topological sector specified by λ the different solutions to the equation $dA = F$ (where F is the harmonic two-form in the topological sector) are in one-to-one correspondence with flat connections. The zero-modes are then collective coordinates for the space of flat connections on X, which is the Jacobian

$$\mathbf{T}^{b_1} = \frac{H^1(X, \mathbf{R})}{H^1(X, \mathbf{Z})}, \tag{8.30}$$

of dimension b_1. This space can be also regarded as the moduli space of Wilson lines of the flat gauge connections. The measure for this space is more subtle. The Jacobi torus (8.30) has a canonical measure which is independent of τ. However, when one considers the non-zero-modes of the gauge field, one has to introduce a factor $(\operatorname{Im} \tau)^{-1/2}$ for each non-zero-mode. Let us denote by B_1 and B_0 the number of one-forms and zero-forms on X. These numbers are in principle infinite, but we could make them finite through a lattice regularization. The number of non-zero-modes would then be

$$B_1 - B_0 + 1 - b_1. \tag{8.31}$$

Here we have subtracted from B_1 the number of non-zero-modes which are pure gauge. There are $B_0 - 1$ of these, since the constant mode acts trivially on the gauge connection. We have also subtracted the number of zero-modes, which is b_1. This gives in principle a factor

$$(\operatorname{Im}\tau)^{\frac{1}{2}(b_1-1)}(\operatorname{Im}\tau)^{\frac{1}{2}(B_0-B_1)}. \tag{8.32}$$

In a local regularization of the theory one should eliminate the second factor. The final result for the measure is then

$$(\operatorname{Im}\tau)^{\frac{1}{2}(b_1-1)}\prod_{i=1}^{b_1} dA_i, \tag{8.33}$$

where A_i are coordinates on the Jacobian (8.30). We can now write the total measure by putting together (8.28) and (8.33) to obtain

$$\frac{1}{(\operatorname{Im}\tau)^{1/2}}\, da\, d\bar{a}\, d\eta_0\, d\chi_0 \prod_{i=1}^{b_1} dA_i dc_i. \tag{8.34}$$

The measure for the commuting and anticommuting collective coordinates for the Jacobi torus simply gives the usual measure for integrating differential forms on $\mathbf{T}^{b_1}$ (this is again a manifestation of the property that functions on superspace being equivalent to differential forms). We can then consider the c_i as a basis of one-forms $\beta_i^\sharp \in H^1(\mathbf{T}^{b_1}, \mathbf{Z})$, dual to $\beta_i \in H^1(X, \mathbf{Z})$. We can then write the zero-modes of ψ as

$$\psi = \sum_{i=1}^{b_1} \beta_i \otimes \beta_i^\sharp. \tag{8.35}$$

Finally, notice that the above identification leaves room for a factor C^{b_1}, which comes from the different normalization of the measures for the b_1 zero-modes between the mathematical definition of Donaldson invariants and the one coming out from physics. It turns out that:

$$C = 2^{9/4}\mathrm{e}^{\pi i/2}. \tag{8.36}$$

8.4. Final form for the u-plane integral

We are now ready to perform the integral over the fermionic zero-modes and the auxiliary fields. First of all, notice that the kinetic terms for the gauge field give the term

$$\exp\left(-i\pi\bar{\tau}(\lambda_+)^2 - i\pi\tau(\lambda_-)^2\right),\qquad(8.37)$$

which, after summing over topological sectors, gives a generalized theta function

$$\vartheta = \sum_{\lambda\in\Gamma+\frac{1}{2}w_2(V)} e^{\left(-i\pi\bar{\tau}(\lambda_+)^2 - i\pi\tau(\lambda_-)^2\right)},\qquad(8.38)$$

where $\Gamma = H^2(X,\mathbf{Z})$. This is a special case of a family of generalized theta functions sometimes called *Siegel–Narain theta functions*, which are characterized by a dependence on $\bar{\tau}$ and show up in Narain compactifications of heterotic string theory. Notice that if we put $w_2(V) = 0$ and we decompose the lattice vectors as

$$\lambda = \sum_{i=1}^{b_2} m_i e_i,\qquad(8.39)$$

where m_i are integers and $\{e_i\}_{i=1,\dots,b_2}$ is a basis of $H^2(X,\mathbf{Z})$, then (8.38) is of the form

$$\sum_{\vec{m}\in\mathbf{Z}^n} g(m),\qquad(8.40)$$

where the function g is a Gaussian involving the quadratic forms

$$P_{ij}^\pm = (e_i,e_j)_\pm\qquad(8.41)$$

which are projectors onto $H^{2,\pm}(X,\mathbf{R})$. Now using Poisson resummation formula,

$$\sum_{\vec{m}\in\mathbf{Z}^n} g(m) = \sum_{\vec{m}\in\mathbf{Z}^n} \hat{g}(m),\qquad(8.42)$$

where

$$\hat{g}(y) = \int d^n x e^{-2\pi i x\cdot y} g(x)\qquad(8.43)$$

is the Fourier transform of g, one can see that (8.38) (for $w_2(V) = 0$) has weight (b_2^-, b_2^+) under $\mathrm{Sl}(2,\mathbf{Z})$ transformations.

In order to soak up the zero-modes of χ and η we have to bring down from the action the vertex

$$-\frac{i\sqrt{2}}{16\pi}\int_X \frac{d\bar{\tau}}{d\bar{a}}\eta\wedge\chi\wedge(F_+ + D_+).\qquad(8.44)$$

Now we integrate out the auxiliary field D, which has a Gaussian action with linear term

$$-\frac{i}{4\pi}\frac{du}{da}\left[(4\pi\lambda_- + D_+)\wedge \widetilde{S}\right] \tag{8.45}$$

where

$$\widetilde{S} = S - \frac{\sqrt{2}}{32}\frac{d\tau}{du}\psi\wedge\psi. \tag{8.46}$$

We now integrate out the D field. Since this is a Gaussian field with the wrong sign, we have to introduce an extra factor of $-i$. The vertex becomes

$$-\frac{\sqrt{2}}{16\pi}\int_X \frac{d\bar{\tau}}{da}\eta\wedge\chi\wedge\left(F_+ + i\frac{(du/da)}{\mathrm{Im}\tau}\widetilde{S}_+\right), \tag{8.47}$$

and we also obtain

$$\exp\left(\widetilde{S}_+^2\left(\frac{(du/da)^2}{8\pi\mathrm{Im}\,\tau}\right)\right). \tag{8.48}$$

Now we can integrate the zero-modes in the vertex insertion. Since F_+ coincides with $4\pi\lambda_+$, and $\chi = \chi_0\omega$, we find an overall factor

$$-\frac{\sqrt{2}}{4}\frac{d\bar{\tau}}{da}\cdot\exp\left(\widetilde{S}_+^2\left(\frac{(du/da)^2}{8\pi y}\right)\right)\cdot\left((\lambda,\omega) + \frac{i}{4\pi y}\frac{du}{da}(\omega,S)\right), \tag{8.49}$$

where $\tau = x + iy$.

Before writing the final answer we have to remember that the effective action includes an extra coupling to the 'gravitational' background given by the second Stiefel–Whitney class of the four-manifold (8.15). In addition, there is an overall phase which encodes the dependence of the Donaldson invariants on a choice of orientation of the moduli space. As we mentioned in Chapter 2, this is specified by a choice of a lifting of $w_2(V)$ to $H^2(X,\mathbf{Z})$, which we will denote by $w = 2\lambda_0$, where $\lambda_0 \in \frac{1}{2}w_2(V)+\Gamma$. This overall sign, together with (8.15), gives the phase

$$(-1)^{(\lambda-\lambda_0)\cdot w_2(X)}e^{2\pi i\lambda_0^2}. \tag{8.50}$$

There is no canonical choice of λ_0 (unless $w_2(V) = 0$, in which case one takes $\lambda_0 = 0$). If λ_0 is replaced by $\widetilde{\lambda}_0$ then (8.50) is multiplied by

$$(-1)^{\beta\cdot w_2(X)} \tag{8.51}$$

where β is the integral class $\beta = \lambda_0 - \widetilde{\lambda}_0$. Thus with the factor (8.50) included the overall sign of the Donaldson invariants depends on a choice of λ_0, and a change of lifting gives a sign (8.51) which agrees with (2.58).

If we now define the following lattice sum:

$$\Psi = \exp\Big[-\frac{1}{8\pi y}\Big(\frac{du}{da}\Big)^2\widetilde{S}_-^2\Big]e^{2\pi i\lambda_0^2}$$

$$\times \sum_{\lambda\in H^2+\frac{1}{2}w_2(V)}(-1)^{(\lambda-\lambda_0)\cdot w_2(X)}\Big[(\lambda,\omega)+\frac{i}{4\pi y}\frac{du}{da}(\widetilde{S},\omega)\Big] \qquad (8.52)$$

$$\times\exp\Big[-i\pi\bar{\tau}(\lambda_+)^2-i\pi\tau(\lambda_-)^2-i\frac{du}{da}(\widetilde{S},\lambda_-)\Big],$$

where the sum over λ is the sum over topological sectors, we see that the final result of the integration is

$$-\frac{\sqrt{2}}{4}\frac{d\bar{\tau}}{d\bar{a}}y^{-1/2}\exp\Big(\widetilde{S}^2\frac{(du/da)^2}{8\pi y}\Big)\cdot\Psi. \qquad (8.53)$$

Combining all the ingredients, we obtain the following expression for the u-plane integral:

$$Z_u = \int_{\Gamma^0(4)\backslash\mathcal{H}}\frac{dxdy}{y^{1/2}}\mu(\tau)$$

$$\int_{\mathbf{T}^{b_1}}\exp\Big[2pu+\widetilde{S}^2\widehat{T}(u)+H(u)(\widetilde{S},\psi^2)+\frac{a_1}{4\sqrt{2}}\frac{du}{da}\int_\gamma\psi+K(u)\psi^4\Big]\Psi(\widetilde{S}), \qquad (8.54)$$

where

$$\mu(\tau) = -\frac{\sqrt{2}}{2}\frac{da}{d\tau}A^\chi B^\sigma C^{b_1},$$

$$\Psi(\widetilde{S}) = \exp(2i\pi\lambda_0^2)\exp\Big[-\frac{1}{8\pi y}\Big(\frac{du}{da}\Big)^2\widetilde{S}_-^2\Big]$$

$$\times\sum_{\lambda\in H^2+\frac{1}{2}w_2(E)}\exp\Big[-i\pi\bar{\tau}(\lambda_+)^2-i\pi\tau(\lambda_-)^2+\pi i(\lambda-\lambda_0,w_2(X))\Big]$$

$$\times\exp\Big[-i\frac{du}{da}(\widetilde{S}_-,\lambda_-)\Big]\Big[(\lambda_+,\omega)+\frac{i}{4\pi y}\frac{du}{da}(\widetilde{S}_+,\omega)\Big],$$

$$\widehat{T}(u) = T(u)+\frac{1}{8\pi\mathrm{Im}\,\tau}\Big(\frac{du}{da}\Big)^2,$$

$$H(u) = \frac{i\sqrt{2}}{64\pi}\Big(\frac{d^2u}{da^2}-4\pi i\frac{d\tau}{du}T(u)\Big),$$

$$K(u) = -\frac{i}{3\cdot 2^{11}\pi}\Big(\frac{d^2\tau}{da^2}-6\frac{d\tau}{du}\frac{d^2u}{da^2}+12\pi i\Big(\frac{d\tau}{du}\Big)^2 T(u)\Big).$$

$$(8.55)$$

In (8.54) the integral over the u-plane is written as an integral over the τ parameter. As discussed in the previous chapter, the moduli space of vacua of $SU(2)$, $\mathcal{N} = 2$ supersymmetric Yang–Mills theory in terms of τ is $\Gamma^0(4)\backslash\mathcal{H}$ given in (7.98), and this is therefore the domain of integration in (8.54). Also notice that ψ is a differential form on the Jacobi torus (8.30), and (8.54) includes an integration over $\mathbf{T}^{b_1}$ coming from the integration over zero-modes that we analysed in the previous section.

Although the detailed properties of the integrand of (8.54) under $\mathrm{Sl}(2, \mathbf{Z})$ transformations will be explained in the next section, we can already use them to determine the contact term T, and therefore the unknown functions $H(u)$, $K(u)$. If ψ has weight $(1,0)$ under $\mathrm{Sl}(2, \mathbf{Z})$ transformations, it follows from (8.46) that $\widetilde{S}$ has weight $(0,0)$. On the other hand, in order to have a well defined behavior under $\mathrm{Sl}(2, \mathbf{Z})$, all the summands in the exponent of (8.54) should have the same weight, which we can read from the first term $2pu$ and is therefore $(0,0)$. It follows that $\widehat{T}(u)$ must have weight $(0,0)$ as well. Let us explore the consequences of this property. Under an S transformation the term

$$G(u) = \frac{1}{8\pi\mathrm{Im}\,\tau}\left(\frac{du}{da}\right)^2 \tag{8.56}$$

in $\widehat{T}(u)$ transforms inhomogeneously:

$$G(u) \to G(u) - \frac{i}{4\pi\tau}\left(\frac{du}{da}\right)^2. \tag{8.57}$$

It follows that T must also transform inhomogeneously, but with the opposite sign, in such a way that $\widehat{T}$ is truly invariant. Under $\tau \to -1/\tau$ one must then have

$$T \to T + \frac{i}{4\pi\tau}\left(\frac{du}{da}\right)^2. \tag{8.58}$$

A comparison with the standard transformation law for the Eisenstein series $E_2(\tau)$ shows that these conditions are equivalent to the statement that

$$T = -\frac{1}{24}E_2(\tau)\left(\frac{du}{da}\right)^2 + H(u), \tag{8.59}$$

where H is modular invariant and so is an ordinary holomorphic function of u. However, there is an extra constraint that $T(u)$ must satisfy: $T(u)$ vanishes at tree level (since it is associated with quantum corrections), therefore it must vanish in the semi-classical region $u \to \infty$. From (7.84) it follows that in this region $du/da \sim 2\sqrt{2u}$, therefore $H(u) = u/3$ and we obtain the final expression for the contact term:

$$T = -\frac{1}{24}\left(E_2(\tau)\left(\frac{du}{da}\right)^2 - 8u\right). \tag{8.60}$$

Using now the explicit expression for $T(u)$, one can easily show that $H(u)$ and $K(u)$ have weight $(-2,0)$ and $(-4,0)$, respectively. Since ψ has weight $(1,0)$ we see that all the terms in the exponent of (8.54) have weight $(0,0)$, as required. We can also obtain the explicit expressions for $H(u)$ and $K(u)$ as follows. Define the functions $f_{1,2}$ by the following equations:

$$\frac{d^2u}{da^2} = 4f_1(q), \quad \frac{d\tau}{da} = \frac{16i}{\pi}f_2(q). \tag{8.61}$$

They have the explicit expression

$$f_1(q) = \frac{2E_2 + \vartheta_2^4 + \vartheta_3^4}{3\vartheta_4^8} = 1 + 24q^{1/2} + \cdots,$$

$$f_2(q) = \frac{\vartheta_2\vartheta_3}{2\vartheta_4^8} = q^{1/8} + 18q^{5/8} + \cdots. \tag{8.62}$$

Using the above result we find that

$$H(u) = \frac{\sqrt{2}}{32}u\frac{d\tau}{du} = \frac{i\sqrt{2}}{16\pi}\frac{\vartheta_2^4 + \vartheta_3^4}{\vartheta_4^8},$$

$$K(u) = \frac{7}{3 \cdot 2^{10}}u\left(\frac{d\tau}{du}\right)^2 = -\frac{7}{3 \cdot 2^7\pi^2}\frac{(\vartheta_2\vartheta_3)^2(\vartheta_2^4 + \vartheta_3^4)}{\vartheta_4^{16}}. \tag{8.63}$$

As a last step we will write (8.54) in a more compact form in the case of manifolds with b_1 even. We explained before that the integral (8.54) is both an integral over the fundamental domain of $\Gamma^0(4)$ and an integral of a differential form over the Jacobi torus $\mathbf{T}^{b_1}$ (remember that ψ is a one-form on this torus). This last aspect can be incorporated in a much more convenient way by using the following simple mathematical results: first, on a manifold of $b_2^+ = 1$, for any β_1, β_2, β_3 and β_4 in $H^1(X, \mathbf{Z})$ one has $\beta_1 \wedge \beta_2 \wedge \beta_3 \wedge \beta_4 = 0$ (this means, in particular, that the last term in the exponent of (8.54) vanishes); second, the image of the map

$$\wedge : H^1(X, \mathbf{Z}) \otimes H^1(X, \mathbf{Z}) \longrightarrow H^2(X, \mathbf{Z}) \tag{8.64}$$

is generated by a single rational cohomology class Λ. We introduce now the antisymmetric matrix a_{ij} associated to the basis β_i of $H^1(X, \mathbf{Z})$, $i = 1, \ldots, b_1$, as $\beta_i \wedge \beta_j = a_{ij}\Lambda$. Finally, we introduce the two-form on $\mathbf{T}^{b_1}$ as

$$\Omega = \sum_{i<j} a_{ij}\beta_i^\sharp \wedge \beta_j^\sharp, \tag{8.65}$$

which does not depend on the choice of basis. This is a volume element for the torus, hence

$$\text{vol}(\mathbf{T}^{b_1}) = \int_{\mathbf{T}^{b_1}} \frac{\Omega^{b_1/2}}{(b_1/2)!} \, . \tag{8.66}$$

Remember that we are assuming b_1 to be even.

We can now write the u-plane integral in a more convenient way. If we define

$$\delta^{\sharp} = \sum_{i=1}^{b_1} \zeta_i \beta_i^{\sharp} \tag{8.67}$$

as the image of δ in (2.60) under the isomorphism $H_1(X, \mathbf{Z}) \simeq H^1(\mathbf{T}^{b_1}, \mathbf{Z})$, we find the final expression:

$$Z_u = -4\pi i \int_{\Gamma^0(4)\backslash H} \frac{dxdy}{y^{1/2}} \int_{\mathbf{T}^{b_1}} h\widehat{f}(p, \delta, S, \tau, y)\Psi(\widetilde{S}), \tag{8.68}$$

with

$$\widehat{f}(p, \delta, S, \tau, y) = \frac{\sqrt{2}}{64\pi} h^{b_1-3} \vartheta_4^{\sigma} f_2^{-1} e^{2pu+S^2\widehat{T}} \exp\left[2f_1(S, \Lambda)\Omega + ih^{-1}\delta^{\sharp}\right], \tag{8.69}$$

where $f_{1,2}$ were defined in (8.61) and we have denoted

$$h = \frac{da}{du}. \tag{8.70}$$

One can also write $\widetilde{S}$ in a more compact way:

$$\widetilde{S} = S - 16f_2 h(\Lambda \otimes \Omega). \tag{8.71}$$

This gives the final expression of the u-plane integral. We now proceed to study its properties and what are its most immediate mathematical applications.

8.5. Behavior under monodromy and duality

In order to understand better the modular properties of the u-plane integral and to write it in a more convenient way, it is useful to introduce a Siegel–Narain theta function considered by Moore and Witten. The formal definition is as follows. Let Λ be a lattice of signature (b_+, b_-), and let P be a decomposition of $\Lambda \otimes \mathbf{R}$ as a sum of orthogonal subspaces of definite signature:

$$P : \Lambda \otimes \mathbf{R} \cong \mathbf{R}^{b_+,0} \perp \mathbf{R}^{0,b_-} \tag{8.72}$$

Let $P_\pm(\lambda) = \lambda_\pm$ denote the projections onto the two factors. We also write $\lambda = \lambda_+ + \lambda_-$. With our conventions $P_-(\lambda)^2 \leq 0$.

Let $\Lambda + \gamma$ denote a translate of the lattice Λ. We then define the following Siegel–Narain theta function:

$$
\begin{aligned}
\Theta_{\Lambda+\gamma} & (\tau, \alpha, \beta; P, \xi) \\
&\equiv \exp\left[\frac{\pi}{2y}(\xi_+^2 - \xi_-^2)\right] \\
&\quad \times \sum_{\lambda \in \Lambda+\gamma} \exp\left[i\pi\tau(\lambda+\beta)_+^2 + i\pi\bar\tau(\lambda+\beta)_-^2 + 2\pi i(\lambda+\beta, \xi)\right. \\
&\qquad\qquad\qquad \left. - 2\pi i(\lambda + \tfrac{1}{2}\beta, \alpha)\right] \\
&= e^{i\pi(\beta,\alpha)} \exp[\frac{\pi}{2y}(\xi_+^2 - \xi_-^2)] \\
&\quad \times \sum_{\lambda \in \Lambda+\gamma} \exp\left[i\pi\tau(\lambda+\beta)_+^2 + i\pi\bar\tau(\lambda+\beta)_-^2 + 2\pi i(\lambda+\beta, \xi)\right. \\
&\qquad\qquad\qquad \left. - 2\pi i(\lambda + \beta, \alpha)\right].
\end{aligned}
\tag{8.73}
$$

The main transformation law of this theta function is:

$$
\Theta_\Lambda\left(-1/\tau, \alpha, \beta; P, \frac{\xi_+}{\tau} + \frac{\xi_-}{\bar\tau}\right) = \sqrt{\frac{|\Lambda|}{|\Lambda'|}}(-i\tau)^{b_+/2}(i\bar\tau)^{b_-/2}\Theta_{\Lambda'}(\tau, \beta, -\alpha; P, \xi),
\tag{8.74}
$$

where Λ' is the dual lattice. This is a generalization of the property, mentioned above, that (8.38) with $w_2(V) = 0$ has weight (b_2^+, b_2^-) under $\mathrm{Sl}(2, \mathbf{Z})$, and it can also be easily proved by using Poisson resummation formula (8.42). If there is a characteristic vector, call it w_2, such that

$$
(\lambda, \lambda) = (\lambda, w_2) \bmod 2
\tag{8.75}
$$

for all λ, then we have, in addition,

$$
\Theta_\Lambda(\tau + 1, \alpha, \beta; P, \xi) = e^{-i\pi(\beta, w_2)/2}\Theta_\Lambda(\tau, \alpha - \beta - \tfrac{1}{2}w_2, \beta; P, \xi).
\tag{8.76}
$$

Let us write the u-plane integral in terms of this theta function. We denote

$$
\Theta = \kappa^{-(w_2(X), w_2(V))}\Theta_\Gamma(\tau, \tfrac{1}{2}w_2(X), \tfrac{1}{2}w_2(E); P_\omega, \xi)
\tag{8.77}
$$

with $\kappa = e^{2\pi i/8}$ and

$$
\xi = \rho y \frac{d\bar a}{du}\omega + \frac{1}{2\pi}\frac{du}{da}\widetilde{S}_-.
\tag{8.78}
$$

We recall that $\Gamma = H^2(X, \mathbf{Z})$. It is easy to see that ξ has the behavior under $\mathrm{Sl}(2, \mathbf{Z})$ prescribed by (8.74). We now introduce the auxiliary integral $\mathcal{G}(\rho)$

$$\mathcal{G}(\rho) \equiv \int_{\Gamma^0(4)\backslash\mathcal{H}} \frac{dxdy}{y^{3/2}} \widehat{f}(p, S, \tau, y)\overline{\Theta}, \tag{8.79}$$

where $\widehat{f}(p, S, \tau, y)$ is given by (8.69). The integral (8.79) is related to the u-plane integral as follows:

$$Z_u = (\widetilde{S}, \omega)\mathcal{G}(\rho)\Big|_{\rho=0} + 2\frac{d\mathcal{G}}{d\rho}\Big|_{\rho=0}. \tag{8.80}$$

Denote the integrand of (8.79) by $(dxdy/y^2)\mathcal{J}$, where $\mathcal{J} = \widehat{f}\cdot y^{1/2}\overline{\Theta}$. It follows from (7.98) that the fundamental domain of $\Gamma^0(4)$ contains six copies of the fundamental domain $\mathcal{F}$. In order to write the u-plane integral explicitly, we map the integrand in these 6 regions to the domain $\mathcal{F}$ defined in (7.95). In order to do this we have to change variables in the integrand according to the different $\mathrm{Sl}(2, \mathbf{Z})$ transformations involved in (7.98). We then obtain six functions:

$$\begin{aligned}
\mathcal{J}_{(\infty,0)}(\tau) &\equiv \mathcal{J}(\tau), \\
\mathcal{J}_{(\infty,1)}(\tau) &\equiv \mathcal{J}(\tau + 1), \\
\mathcal{J}_{(\infty,2)}(\tau) &\equiv \mathcal{J}(\tau + 2), \\
\mathcal{J}_{(\infty,3)}(\tau) &\equiv \mathcal{J}(\tau + 3), \\
\mathcal{J}_M(\tau) &\equiv \mathcal{J}(-1/\tau), \\
\mathcal{J}_D(\tau) &\equiv \mathcal{J}(2 - 1/\tau).
\end{aligned} \tag{8.81}$$

The subscript D refers to 'dyon', and should not be confused with the subscript for dual quantities. In general we denote $\Gamma^0(4)$-modular forms F transformed as in (8.81) by F_I where

$$I = (\infty, 0), (\infty, 1), (\infty, 2), (\infty, 3), M, D. \tag{8.82}$$

The integral (8.79) then becomes

$$\mathcal{G}(\rho) = \int_{\mathcal{F}} \frac{dxdy}{y^{3/2}} \sum_I \widehat{f}_I(p, S, \tau)\overline{\Theta}_I, \tag{8.83}$$

where

$$\Theta_I = e^{i\phi_I}\Theta(\tau, \alpha_I, \beta_I; \xi_I) \tag{8.84}$$

are the transforms of the Siegel–Narain theta function implied by (8.81), and the ϕ_I are the appropriate phases. One can check that:

$$e^{i\phi_{(\infty,n)}} = \kappa^{-(n+1)(w_2(X),w_2(E))}, \quad n = 0,1,2,3,$$
$$e^{i\phi_M} = e^{i\phi_D} = \kappa^{-(w_2(X),w_2(E))},$$

(8.85)

and that

$$\alpha_{(\infty,0)} = \frac{1}{2}w_2(X), \qquad\qquad \beta_{(\infty,0)} = \frac{1}{2}w_2(V),$$
$$\alpha_{(\infty,1)} = -\frac{1}{2}w_2(V), \qquad\qquad \beta_{(\infty,1)} = \frac{1}{2}w_2(V),$$
$$\alpha_{(\infty,2)} = -w_2(V) - \frac{1}{2}w_2(X), \qquad \beta_{(\infty,2)} = \frac{1}{2}w_2(V),$$
$$\alpha_{(\infty,3)} = -\frac{3}{2}w_2(V) - \frac{1}{2}w_2(X), \qquad \beta_{(\infty,3)} = \frac{1}{2}w_2(V),$$
$$\alpha_M = \frac{1}{2}w_2(V), \qquad\qquad \beta_M = -\frac{1}{2}w_2(X),$$
$$\alpha_D = \frac{1}{2}w_2(V), \qquad\qquad \beta_D = w_2(V) + \frac{1}{2}w_2(X).$$

(8.86)

In order to show that the integrand of (8.83) has weight $(0,0)$ one has to check that $\widehat{f}_I$ and Θ_I are in the same unitary representation of $\mathrm{Sl}(2,\mathbf{Z})$. The behavior of the $\widehat{f}_I$ under the generators T and S is as follows:

$$\begin{aligned}
\widehat{f}_{(\infty,0)}(\tau+1) &= \widehat{f}_{(\infty,1)}(\tau), & \widehat{f}_{(\infty,0)}(-1/\tau) &= (-i\tau)^{\sigma/2}\widehat{f}_M(\tau),\\
\widehat{f}_{(\infty,1)}(\tau+1) &= \widehat{f}_{(\infty,2)}(\tau), & \widehat{f}_{(\infty,1)}(-1/\tau) &= (-i\tau)^{\sigma/2}\widehat{f}_{(\infty,3)}(\tau),\\
\widehat{f}_{(\infty,2)}(\tau+1) &= \widehat{f}_{(\infty,3)}(\tau), & \widehat{f}_{(\infty,2)}(-1/\tau) &= (-i\tau)^{\sigma/2}\widehat{f}_D(\tau),\\
\widehat{f}_{(\infty,3)}(\tau+1) &= \widehat{f}_{(\infty,0)}(\tau), & \widehat{f}_{(\infty,3)}(-1/\tau) &= (-i\tau)^{\sigma/2}\widehat{f}_{(\infty,1)}(\tau),\\
\widehat{f}_M(\tau+1) &= \kappa^\sigma \widehat{f}_M(\tau), & \widehat{f}_M(-1/\tau) &= (-i\tau)^{\sigma/2}\widehat{f}_{(\infty,0)}(\tau),\\
\widehat{f}_D(\tau+1) &= \kappa^\sigma \widehat{f}_D(\tau), & \widehat{f}_D(-1/\tau) &= (-i\tau)^{\sigma/2}\widehat{f}_M(\infty,2)(\tau).
\end{aligned}$$

(8.87)

Using (8.85) and (8.86) it is easy to show that the Θ_I are in the same unitary representation as $\widehat{f}_I$, therefore the integrand is invariant under $\mathrm{Sl}(2,\mathbf{Z})$. To prove these properties it is crucial to take into account the Wu formula (1.22) and also that $w_2(X)^2 \equiv \sigma \bmod 8$, a consequence of (1.27).

As a corollary of the above analysis one can show that for every I $\widehat{f}_I\overline{\Theta}_I$ is invariant under the monodromy associated to the corresponding cusp, as required by consistency. Let us consider, for example, the cusp $(\infty,0)$. The monodromy is given by T^4, which maps $\widehat{f}_{(\infty,0)} \to \widehat{f}_{(\infty,0)}$ and $\Theta_{(\infty,0)} \to \Theta_{(\infty,0)}$,

therefore the product is clearly invariant. If we now look at the monopole
cusp the monodromy in magnetic variables is given by T, and we see that
$\widehat{f}_M$ and Θ_M transform with the same phase, so that $\widehat{f}_M \overline{\Theta}_M$ is invariant.
In this analysis (as well as in the analysis of invariance under $\mathrm{Sl}(2, \mathbf{Z})$), the
presence of the coupling (8.15) is crucial for consistency. In fact, this coupling
is required by invariance under the monodromy at infinity. This provides a
very easy way of deducing the existence of such a coupling from consistency
of the u-plane integral which can also be generalized to more complicated
situations, such as the higher rank case which we will briefly consider in the
last chapter. As a final remark notice that the coupling (8.15) indicates that
the lattice in the monopole cusp is $\Gamma + \frac{1}{2}w_2(X)$. Taking into account (3.13)
we see that the magnetic line bundles are in fact Spin_c structures. This will
be crucial in the analysis of the contributions of $u = \pm 1$.

Bibliographical notes

- The u-plane integral was introduced and developed by Moore and
  Witten in their fundamental work [61]. A short summary can be
  found in [100] and in [101]. A related, but somewhat different,
  point of view of Donaldson–Witten theory can be found in [102].
  The extension to the non-simply connected case is also considered in
  [61][102], and it is worked out in detail in [103][104].

- The appearance of couplings to gravity in four-dimensional topological gauge theories involving the signature and Euler characteristic
  was first pointed out in [105]. In the case of Donaldson–Witten theory the couplings were derived in [106], which also discusses the measure for the zero-modes of the gauge field as well as the appearance
  of (8.15).

- The appearance of contact terms for the observables associated with
  two-cycles is already pointed out in [107], and they are derived in
  [61]. In the text we have only discussed contact terms for observables
  associated to the intersection of two-cycles, but on a general four-
  manifold there are more contact terms associated with intersections
  of the various p-cycles. A general proposal for these contact terms
  was made in [102] and applied in [103] to studying the non-simply
  connected case in full generality. A more conceptual understanding
  of $T(u)$ and its relation to the Seiberg–Witten prepotential can be
  found in [102][100][108].

- The mathematical results discussed in relation with (8.64) are derived in [21].

Chapter 9

Some Applications of the u-plane Integral

In this chapter we will discuss three important applications of the u-plane integral to Donaldson invariants: the wall crossing formula, the blow-up formula, and a precise determination of the Seiberg–Witten contributions at the monopole points. This last application will in fact lead us to Witten's 'magic formula' for the Donaldson series of four-manifolds with $b_2^+ > 1$ and of simple type in terms of Seiberg–Witten invariants. All of these properties can be derived from the form of the integrand, $i.e.$, it is not necessary to perform the integral over the fundamental domain of $\Gamma^0(4)$ explicitly.

9.1. Wall crossing

We have seen that the u-plane integral has good properties with respect to duality transformations and monodromy invariance. However, this does not mean that it is well defined: the integration region is non-compact, and the integrand is ill behaved in general as $\tau \to \infty$. We then have to define the integral carefully, in such a way that it makes sense and that (hopefully) we recover the results of Donaldson theory.

The precise recipe was worked out in detail by Moore and Witten: first, to obtain a Donaldson invariant of some given order we expand Z_u to the required order in p and S. This gives an integral which computes a particular Donaldson invariant. To define that particular integral we write $\tau = x + iy$, we perform the integral for $y < y_0$ for some cutoff y_0, and then take the limit as $y_0 \to \infty$ only at the end. A similar procedure is followed near the cusps at $u = \pm 1$, introducing the dual τ-parameters and integrating first over $\operatorname{Im} \tau_D < y_0$, before taking the limit as $y_0 \to \infty$. This procedure eliminates the infinities. Let us see this in some detail. Set $q = \exp(2\pi i \tau)$. Then the integral to a given order in p and S is a sum of terms, each of which is a power of y multiplied by a sum of the form

$$\sum_{\nu,\mu} q^\nu \bar{q}^\mu. \tag{9.1}$$

Although ν has no lower bound, μ is bounded below by zero. This is because negative exponents in (9.1) come only from factors in the integrand, such as u and $(d\tau/du)^{-\sigma/4}$, which are singular at the cusps. But these factors are holomorphic and so contribute to ν but not μ.

Consider now an integral of the following form:

$$\lim_{y_0 \to \infty} \int_{y_1}^{y_0} \frac{dy}{y^c} \int_0^k dx \sum_{\nu,\mu} q^\nu \bar{q}^\mu, \tag{9.2}$$

where y_1 is an arbitrary lower cutoff. We want to find a prescription to compute this integral in such a way that it converges when $y_0 \to \infty$. The x integral runs from 0 to k where (for $\Gamma^0(4)$) $k = 4$ for the cusp at infinity, and $k = 1$ for the other cusps. If we now look at the integrand and to the definition of the function Ψ, we see that in all cases either $c > 1$ or there are, for a generic metric on X, no terms with $\nu = \mu = 0$. If we now integrate first over x we project the sum in (9.2) onto terms with $\nu = \mu$, and hence (as μ is non-negative) onto terms that vanish exponentially or, if $\nu = \mu = 0$, are constant at infinity. For a generic metric on X, the y integral converges as $y_0 \to \infty$ since all terms that have survived the x integral have $c > 1$ or $\nu, \mu > 0$. With this prescription the integral becomes for a *generic metric* a well defined formal power series in p, S.

For *special metrics* one can find, however, terms with $c = 1/2$ and $\nu = \mu = 0$. The relevant terms are of the form

$$\mathcal{I}(\omega) \equiv \int_{\mathcal{F}} \frac{dxdy}{y^{1/2}} c(d) e^{2\pi i x d - 2\pi y d} e^{-i\pi x(\lambda_+^2 + \lambda_-^2)} e^{-\pi y(\lambda_+^2 - \lambda_-^2)}(\omega, \lambda) \tag{9.3}$$

for some integer d and some λ. In (9.3) $c(d)$ is the coefficient of some modular from. We want to study the integral in (9.3) for fixed λ as the decomposition $\lambda = \lambda_+ + \lambda_-$ varies. Since $\lambda_+ = \omega(\omega, \lambda)$ this decomposition is determined by the period point ω. We want to know if there are discontinuities in $\mathcal{I}(\omega)$ when a 'wall' is crossed. Such a discontinuity occurs only if $\lambda^2 < 0$, since otherwise the integral in (9.3) (with the regularization just described) is convergent. However, when $\lambda^2 < 0$ there is a discontinuity at $\lambda_+ = 0$. This can be computed as follows. Upon performing the x integral one projects onto d such that $2d = \lambda^2$. For this value of d the y integral looks like

$$\int_{y_1}^{\infty} \frac{dy}{y^{1/2}} \, c(\lambda^2/2) e^{-2\pi y \lambda_+^2} \lambda_+. \tag{9.4}$$

This is an elementary integral (if one replaces y_1 by 0) and converges for all non-zero λ_+, but is discontinuous at $\lambda_+ = 0$. The discontinuity comes from the large y part of the integral and so is independent of y_1. The discontinuity in $\mathcal{I}(\omega)$ as ω crosses from $(\omega, \lambda) = 0^-$ to $(\omega, \lambda) = 0^+$ is easily computed to be (for a single copy of the $\mathrm{Sl}(2, \mathbf{Z})$ fundamental domain)

$$\mathcal{I}(\omega_+) - \mathcal{I}(\omega_-) = \sqrt{2} c(d) = \sqrt{2} \left[q^{-\lambda^2/2} c(q) \right]_{q^0}. \tag{9.5}$$

The notation $[\cdot]_{q^0}$ indicates the constant term in a Laurent expansion in powers of q, and it may also be expressed as a residue. Since $\lambda_+ = 0$ we may put $(S_+, \lambda_+) = 0$ and $(S_-, \lambda_-) = (S, \lambda)$ in the function $c(q)$. Note that in order to compute the wall crossing behavior of the u-plane integral we do not have to compute the integral itself, but we only need information about the integrand.

Let us look at the four cusps at infinity. In all cases one has $2\lambda \equiv w_2(V)$ mod 2, and we see that the conditions $\lambda^2 < 0$, $\lambda_+ = 0$ for a discontinuity in the integral are precisely the conditions for wall crossing of the Donaldson invariants that we explained in (2.77) (with $\zeta = 2\lambda$). The general formula (9.5) gives

$$\begin{aligned}
\mathrm{WC}_{\zeta = 2\lambda} = &-\frac{i}{2} (-1)^{(\lambda - \lambda_0, w_2(X))} e^{2\pi i \lambda_0^2} \\
&\times \Big[q^{-\lambda^2/2} h_{\infty}(\tau)^{b_1 - 2} \vartheta_4^{\sigma} f_{2\infty}^{-1} \exp\left(2pu_{\infty} + S^2 T_{\infty} - i(\lambda, S)/h_{\infty}\right) \\
&\times \int_{\mathbf{T}^{b_1}} \exp\left(2f_{1\infty}(q)(S, \Lambda)\Omega + 16i f_{2\infty}(q)(\lambda, \Lambda)\Omega + i\frac{\delta^{\#}}{h_{\infty}}\right) \Big]_{q^0},
\end{aligned} \tag{9.6}$$

where we have included a factor of 4 to take into account the contributions from the four cusps. This expression gives, in fact, the wall crossing behavior for the Donaldson invariants of manifolds with even $b_1 \geq 0$. An obvious corollary of (9.6) is that the wall crossing formula only depends on the cohomology ring of the four-manifold X, confirming the conjecture of Kotschick and Morgan that we mentioned in Chapter 2. Using the behavior of the modular forms involved in the wall crossing formula, it is easy to check that the wall crossing term is different from zero only if

$$0 > \lambda^2 \geq p_1/4, \tag{9.7}$$

where p_1 is the Pontrjagin number of the gauge bundle. As an example of the use of (9.6), consider $\lambda^2 = p_1/4$. The wall crossing term for the Donaldson invariant corresponding to $p^r S^{d-2r}$, where d is half the dimension of moduli space, is easily found to be:

$$\mathrm{WC}_\zeta (p^r S^{d-2r}) = \frac{1}{2}(-1)^{(\lambda - \lambda_0, w_2(X))} 2^{3b_1/2 - b - d}(-1)^{r+d} p^r \mathrm{vol}(\mathbf{T}^{b_1})$$

$$\times \sum_{b=0}^{b_1/2} \frac{(b_1/2)!}{(b_1/2 - b)!} \binom{d-2r}{b} (S, \zeta)^{d-2r-b}(S, \zeta)^b (\zeta, \Lambda)^{b_1/2 - b}, \tag{9.8}$$

where $\zeta = 2\lambda$. For simply connected manifolds the general wall crossing formula (9.6) reproduces the result of Göttsche.

9.2. The Seiberg–Witten contribution

In the first section of this chapter we computed the wall crossing formula only for the cusps at infinity. However, there is also wall crossing at the monopole and the dyon cusps. Let us consider the monopole cusp in some detail (the results for the dyon cusp can be obtained in a very similar way). The first thing to notice is that after performing the S-duality transformation the lattice points λ live in $H^2(X, \mathbf{Z}) + w_2(X)/2$. This means that 2λ corresponds to a Spin_c-structure. The walls are still defined by

$$\lambda^2 < 0, \quad (\lambda, \omega) = 0. \tag{9.9}$$

and the wall crossing term is easily computed as:

$$\frac{i}{8} e^{2i\pi(\lambda_0 \cdot \lambda + \lambda_0^2)} \left[q_D^{-\lambda^2/2} h_M^{b_1 - 3} \vartheta_2^{8+\sigma} \exp\left(2pu_M - i(S, \lambda)/h_M + S^2 T_M(u)\right) \right.$$

$$\left. \times \int_{\mathbf{T}^{b_1}} \exp\left(2f_{1M}(S, \Lambda)\Omega + 16if_{2M}(\lambda, \Lambda)\Omega + \frac{i}{h_M}\delta^\sharp\right) \right]_{q_D^0}. \tag{9.10}$$

where the 'dual' modular forms are explicitly given by

$$u_M(q_D) = \frac{\vartheta_3^4 + \vartheta_4^4}{2(\vartheta_3\vartheta_4)^2} = 1 + 32q_D + 256q_D + \cdots,$$

$$h_M(q_D) = \frac{1}{2i}\vartheta_3\vartheta_4 = \frac{1}{2i}(1 - 4q_D + 4q_D^2 + \cdots),$$

$$f_{1M}(q_D) = \frac{2E_2 - \vartheta_3^4 - \vartheta_4^4}{3\vartheta_2^8} = -\frac{1}{8}(1 - 6q_D + 24q_D^2 + \cdots), \qquad (9.11)$$

$$f_{2M}(q_D) = \frac{\vartheta_3\vartheta_4}{2i\vartheta_2^8} = \frac{1}{2^9 i}(\frac{1}{q_D} - 12 + 72q_D + \cdots),$$

$$T_M(q_D) = -\frac{1}{24}\left(\frac{E_2}{h_M^2} - 8u_M\right) = \frac{1}{2} + 8q_D + 48q_D^2 + \cdots,$$

and $q_D = \exp(2\pi i \tau_D)$.

After computing the leading power of q_D in (3.29) for a wall given by λ, one finds that monopole wall crossing only takes place if

$$d_\lambda \geq 0, \qquad (9.12)$$

where d_λ, given in (3.14), is the dimension of the Seiberg–Witten moduli space associated to the Spin$_c$-structure with determinant line bundle 2λ. Therefore the conditions for a discontinuity of the u-plane integral at the monopole cusp (9.9) and (9.12) are precisely the conditions for wall crossing in the Seiberg–Witten invariants (3.29). The picture that emerges from this analysis is the following. First, let us recall that according to the basic principle 'Donaldson = Coulomb+Higgs', the Donaldson–Witten generating function is a sum of the three terms in (8.1), which we reproduce here:

$$Z_{\mathrm{DW}} = Z_u + Z_{u=1} + Z_{u=-1}. \qquad (9.13)$$

When there is a Donaldson wall the u-plane integral jumps as a result of to the behavior at the cusps at infinity. This jump gives the Donaldson wall crossing, as we saw in the previous section. When there is a Seiberg–Witten wall, Z_u jumps at the monopole and dyon cusps, but since Z_{DW} has no wall crossing at these walls (generically), the jump should be compensated by an identical jump but with opposite sign in $Z_{u=\pm1}$. Since, as we will see in a moment, $Z_{u=\pm1}$ involves the Seiberg–Witten invariants, in order to have a cancellation of wall crossings we should have the same conditions for the walls. Happily, and as we have just seen, this turns out to be the case.

In fact, more is true: as Moore and Witten realized, cancellation of wall crossings can be used to determine the detailed form of $Z_{u=\pm 1}$. Let us now focus on the analysis of the contribution at $u = 1$ (the analysis of the contribution at $u = -1$ is very similar and, in fact, can be obtained from the $\mathbf{Z}_2$ symmetry of the u-plane). The first thing to notice is that since at $u = 1$ there is a preferred frame (the magnetic frame obtained from S-duality from the electric frame suitable for the semi-classical region) we can perform an analysis of the functional integral based on localization on supersymmetric configurations, as we did in Chapter 5 for Donaldson–Witten theory and the monopole theory. We will then start with the twisted Lagrangian of the theory. We saw in Chapter 7 that the effective Lagrangian of the physical theory near $u = 1$ involves both the $\mathcal{N} = 2$ supersymmetric vector multiplet (in the dual frame) and the monopole hypermultiplet, and involves the magnetic prepotential $\widetilde{\mathcal{F}}_D(a_D)$. It is easy to see that, after twisting, this Lagrangian can be written as:

$$
\begin{aligned}
\{\overline{\mathcal{Q}}, W\} &+ \frac{i}{16\pi}\tilde{\tau}_D F \wedge F + p(u)\mathrm{Tr}\, R \wedge *R + \ell(u)\mathrm{Tr} R \wedge R \\
&- \frac{i\sqrt{2}}{32 \cdot \pi}\frac{d\tilde{\tau}_D}{da_D}(\psi \wedge \psi) \wedge F + \frac{i}{3 \cdot 2^7 \pi}\frac{d^2\tilde{\tau}_D}{da_D^2}\psi \wedge \psi \wedge \psi \wedge \psi.
\end{aligned}
\tag{9.14}
$$

We have included in this Lagrangian couplings to gravity similar to those considered in our analysis of the u-plane integral in the previous chapter, since on general grounds we should expect them on a curved manifold. The part of the Lagrangian involving the monopole hypermultiplet can also be written as a $\overline{\mathcal{Q}}$-exact term after twisting (this is essentially the matter piece in (5.57)), and has been included in W. Our normalizations are such that M, the monopole field, is a section of $L^{1/2}$ (the dual line bundle), and then F is the curvature of L (this can be checked by taking into account the rescaling (7.60), which also explains why some of the factors differ from (8.2)). If $\lambda = c_1(L)$ we then have $F = 4\pi\lambda$. The exponentiation of the terms involving the densities $\mathrm{Tr}\, R \wedge *R$, $\mathrm{Tr}\, R \wedge R$ and λ^2 gives after integration on X the factors

$$
P(u)^{\sigma/8}L(u)^{\chi/4}C(u)^{\lambda^2/2},
\tag{9.15}
$$

where $C(u) = \mathrm{e}^{-2\pi i\tilde{\tau}_D}$. As in the u-plane integral, in the monopole theory the ψ field is a one-form on the Seiberg–Witten moduli space, and a one-form on the four-manifold X. We can then write it as

$$\psi = \frac{c}{2} \sum_{i=1}^{b_1} \nu_i \beta_i, \qquad (9.16)$$

where $\beta_i \in H^1(X, \mathbf{Z})$, $i = 1, \ldots, b_1$, is the basis of one-forms considered in (8.27), the ν_i are the one-forms on moduli space defined in (3.20), and c is a normalization constant (in order to agree with mathematical normalizations, one has to put $c = 2^{-9/4} \pi^{-1/2} i$).

Finally, we have to consider how to insert observables in this theory. This is just as in the u-plane theory, by using the descent procedure. Therefore

$$p\mathcal{O} \to 2pu_M, \qquad (9.17)$$

and

$$\exp(I_2(S)) \to \exp\left(-\frac{i}{4\pi} \int_S \frac{du}{da_D} F + S^2 T_M(u) + \frac{i}{8\sqrt{2}\pi} \frac{d^2u}{da_D^2} \int_S \psi \wedge \psi\right). \qquad (9.18)$$

Notice that in the $U(1)$ theory coupled to matter one has $\{G_\mu, \psi_\nu\} = -(F_{\mu\nu}^- + D_{\mu\nu}^+) + 2i(\overline{M}M)_{\mu\nu}^+$, but it also follows from (5.56) that supersymmetric configurations satisfy $F_{\dot\alpha\dot\beta}^+ + 2i\overline{M}_{(\dot\alpha} M_{\dot\beta)} = D_{\alpha\dot\beta}$. Therefore when considering supersymmetric configurations one finds indeed (9.18). Notice that since a_D is invariant under the rescaling (7.60), we have to use the original G action in (5.32) (in contrast to (8.19)). Finally, we also have:

$$I_1(\delta) \to \frac{\sqrt{2}a_1}{4} \frac{1}{h_M} \int_\delta \psi. \qquad (9.19)$$

Let us now evaluate the contribution from the monopole theory. As we discussed in Chapter 5, the functional integral reduces to an integral over the space of collective coordinates for the supersymmetric configurations. This is precisely the Seiberg–Witten moduli space. The observables, together with the measure factors and the terms in the Lagrangian that are not $\overline{Q}$-exact, give a function of a_D and the ν_i that one then integrates over this moduli space. a_D is nothing but the operator ϕ in (3.19), and ψ gives insertions of ν_i operators. The contribution to $Z_{u=1,\lambda}$ from the Spin$_c$-structure associated to λ is then given by:

$$Z_{u=1,\lambda} = \int_{\mathcal{M}_\lambda} 2e^{2i\pi(\lambda_0^2 - \lambda_0 \cdot \lambda)} C(u)^{\lambda^2/2} P(u)^{\sigma/8} L(u)^{\chi/4}$$

$$\times \exp\left(2pu_M + i(S, \lambda)/h_M + S^2 T_M(u)\right) \qquad (9.20)$$

$$\times \exp\left[c^2(P_M(u), \Lambda) \sum_{i,j=1}^{b_1} a_{ij} \nu_i \nu_j\right],$$

where

$$P_M(u) = \frac{i\sqrt{2}}{16\pi} f_{1M} S + \frac{i\sqrt{2}}{32} \frac{d\tilde{\tau}_D}{da_D} \lambda, \qquad (9.21)$$

and we have taken into account that $da_D/du = -h_M$. The factor of 2 comes from the same normalization issue that we discussed in (8.24), and for simplicity we have put $I_1(\delta) = 0$. The overall phase depending on λ_0 can be obtained by performing carefully the duality transformation of the effective Lagrangian in the presence of a non-trivial Stiefel–Whitney class. As we will see in a moment, it is clearly required to match the wall crossings.

The integral (9.20) can be evaluated in terms of Seiberg–Witten invariants by expanding the exponential, and extracting the different powers of a_D and the ν_i. By doing this we find:

$$
\begin{aligned}
Z_{u=1,\lambda} = \sum_{b=0}^{b_1/2} \frac{1}{b!} \mathrm{Res}_{a_D=0} &\Big[2e^{2i\pi(\lambda_0^2 - \lambda_0 \cdot \lambda)} C(u)^{\lambda^2/2} P(u)^{\sigma/8} L(u)^{\chi/4} \\
&\times \exp\left(2pu_M + i(S,\lambda)/h_M + S^2 T_M(u) \right) \\
&\times a_D^{-d_\lambda/2 + b - 1} (P_M(u), \Lambda)^b \Big] \\
&\times c^{2b} \sum_{i_p, j_p = 1}^{b_1} a_{i_1 j_1} \cdots a_{i_b j_b} \mathrm{SW}_\lambda(\beta_{i_1} \wedge \beta_{j_1} \wedge \cdots \wedge \beta_{i_b} \wedge \beta_{j_b}).
\end{aligned}
$$

$$(9.22)$$

Let us now compare to (3.29) (and use the property that (3.29) is odd under the change $\lambda \to -\lambda$). We see that in order to match (3.29) and (9.22) consistency requires that λ lives in the lattice $H^2(X, \mathbf{Z}) + w_2(X)/2$, so the dual line bundle $L^{1/2}$ is, in fact, a Spin$_c$ structure. The reason behind this is again the crucial coupling (8.15), since after the S-duality transformation it generates the appropriate shift in the dual λ. The effect of (8.15) in the dual theory can be analysed directly, and leads to the same conclusion.

We can now determine the unknown factors in (9.22) by requiring that the wall crossing of the above expression matches the wall crossing at the monopole cusp. This is easily done by considering first the simply connected case $b_1 = 0$, since in that case we know that the Seiberg–Witten invariant jumps by ± 1 in crossing a wall. The Seiberg–Witten wall crossing (9.10) can be also written as a residue in order to compare with (9.22), and in the simply connected case it is given by

$$\text{Res}_{a_D=0} \left[4\pi q_D^{-\lambda^2/2} \alpha^\chi \left(\frac{du}{da_D} \right)^{\chi/2} \beta^\sigma (u^2 - 1)^{\sigma/8} \right.$$

$$\left. \times \exp \left(2pu_M + i(S,\lambda)/h_M + S^2 T_M(u) \right) \right].$$

(9.23)

For simply connected manifolds with $b_2^+ = 1$ we can write $\chi = 4 - \sigma$, and by comparing the terms involving σ and λ^2, one finds:

$$C = \frac{a_D}{q_D},$$

$$P = -\pi^2 \beta^8 a_D^{-1}(u^2 - 1),$$

$$L = \pi i \alpha^4 \left(\frac{du}{da_D} \right).$$

(9.24)

In terms of modular forms and taking into account the values of the constants α, β given in (8.17), we have

$$P = -\frac{4\vartheta_2(\tau_D)^8}{h_M^4} a_D^{-1}, \quad L = \frac{i}{h_M^2}.$$

(9.25)

The first relation tells us that the gauge coupling $\tilde{\tau}_D$ appearing in (9.14) is given by

$$\tilde{\tau}_D = \tau_D - \frac{1}{2\pi i} \log a_D,$$

(9.26)

and therefore it is smooth at the monopole cusp. This defines the prepotential $\tilde{\mathcal{F}}_D(a_D)$ through the equation $\tilde{\mathcal{F}}_D''(a_D) = \tilde{\tau}_D$, and finally proves our claim in (7.94). With this information we can already compute the remaining couplings in (9.14). Since

$$\frac{d\tau_D}{da_D} = -f_{2M}(q_D), \quad a_D = -\frac{i}{6} \frac{2E_2(\tau_D) - \vartheta_3^4 - \vartheta_4^4}{\vartheta_3 \vartheta_4} = -\frac{1}{4} \frac{f_{1M}}{f_{2M}}$$

(9.27)

one finds

$$P_M(u) = \frac{i\sqrt{2}}{32\pi} \left[2f_{1M}S - 16i \left(\frac{1 + 8f_{1M}}{8f_{1M}} \right) f_{2M}\lambda \right].$$

(9.28)

One can actually require cancellation of the wall crossings in the non-simply connected case and derive the general formula for Seiberg–Witten wall crossing. For example, when no insertions of the ν_i are made one finds

$$\text{WC}_\lambda = (-1)^{b_1/2}(\lambda, \Lambda)^{b_1/2} \text{vol}(\mathbf{T}^{b_1}),$$

(9.29)

in agreement with known mathematical results.

This gives in principle all the information which is needed in order to write the Seiberg–Witten contribution at $u = 1$ to the Donaldson invariants, for any four-manifold with $b_2^+ \geq 1$. For example, for a manifold with $b_2^+ = 1$, and after including $I_1(\delta)$, one obtains:

$$Z_{u=1,\lambda} = \frac{i^{b_1+1}}{8} \sum_{b \geq 0} \sum_{n=0}^{b} \frac{(-1)^n}{n!(b-n)!} 2^{-6n-5b+b_1/2} e^{2i\pi(\lambda_0^2 - \lambda_0 \cdot \lambda)}$$

$$\times \left[q_D^{-\lambda^2/2} h_M^{b_1-3} \vartheta_2^{8+\sigma} \left(\frac{f_{1M}}{if_{2M}} \right)^{(b+n-b_1)/2} \right.$$

$$\exp\left(2pu_M + i(S,\lambda)/h_M + S^2 T_M(u)\right)$$

$$\left(2f_{1M}(S,\Lambda) - 16i \left(\frac{1+8f_{1M}}{8f_{1M}} \right) f_{2M}(\lambda,\Lambda) \right)^n \Bigg]_{q_D^0}$$

$$\times \sum_{i_p,j_p=1}^{b_1} a_{i_1 j_1} \cdots a_{i_n j_n} \mathrm{SW}_\lambda(\beta_{i_1} \wedge \beta_{j_1} \wedge \cdots \wedge \beta_{i_n} \wedge \beta_{j_n} \wedge \delta_*^{b-n}),$$

$$(9.30)$$

where δ_* denotes the image of δ under the map $\delta_i \to \beta_i$, and we have taken into account that on manifolds with $b_2^+ = 1$ the term involving ψ^4 in (9.14) vanishes (as we mentioned in the previous chapter).

The contribution $Z_{u=-1}$ at $u = -1$ is identical to the contribution at $u = 1$, with the only difference that one has to use the modular forms

$$u_D = -u_M, \qquad h_D = ih_M, \qquad f_{1D} = f_{1M}, \qquad f_{2D} = if_{2M}, \qquad T_D = -T_M$$

$$(9.31)$$

and include an extra factor $\exp(-2\pi i \lambda_0^2)$. It is easy to check that

$$Z_{u=-1}(p, \zeta_i, v_i) = e^{-2\pi i \lambda_0^2} i^{(\chi+\sigma)/4} Z_{u=1}(-p, i\zeta_i, -iv_i). \qquad (9.32)$$

For a general four-manifold the Seiberg–Witten contributions at $u = \pm 1$ are rather complicated. For example, in a manifold with $b_2^+ > 1$, $b_1 > 0$ and which is not of simple type, the term ψ^4 in (9.14) gives a contribution. For a simply connected manifold (but not necessarily of Seiberg–Witten simple type) the above expressions give:

$$Z_{u=1,\lambda} = \frac{\mathrm{SW}(\lambda)}{16} \cdot e^{2i\pi(\lambda_0^2 - \lambda_0 \cdot \lambda)}$$

$$\times \left[q_D^{-\lambda^2/2} \frac{\vartheta_2^{8+\sigma}}{a_D h_M} \left(-2i \frac{a_D}{h_M^2} \right)^{\chi_h} \exp\left(2pu_M + i(\lambda,S)/h_M + S^2 T_M\right) \right]_{q^0}.$$

$$(9.33)$$

If the manifold has $b_2^+ > 1$ and is of Seiberg–Witten simple type, the Seiberg–Witten contributions simplify enormously. Since $Z_u = 0$ the Donaldson–Witten generating function is given by $Z_{\mathrm{DW}} = Z_{u=1} + Z_{u=-1}$. The Seiberg–Witten simple type condition means that the only contribution comes from basic classes λ with $d_\lambda = 0$, therefore we only have to take the leading terms in the q-expansions in (9.33). Using $u_M = 1 + \cdots$, $T_M = 1/2 + \cdots$, $h_M = 1/(2i) + \cdots$ and $a_D = 16iq_D + \cdots$ we find that (9.33) reduces to

$$(-1)^{\chi_h} 2^{1 + \frac{7\chi}{4} + \frac{11\sigma}{4}} e^{2p + S^2/2} e^{-2(S,\lambda)} e^{2i\pi(\lambda_0^2 - \lambda_0 \cdot \lambda)} \mathrm{SW}(\lambda). \qquad (9.34)$$

We then obtain, after summing over all λ and taking into account (3.26) and (9.32),

$$Z_{\mathrm{DW}} = 2^{1 + \frac{7\chi}{4} + \frac{11\sigma}{4}} \sum_\lambda e^{2i\pi(\lambda_0 \cdot \lambda + \lambda_0^2)} \left[e^{2p + S^2/2} e^{2(S,\lambda)} \right.$$
$$\left. + i^{\chi_h - w_2(V)^2} e^{-2p - S^2/2} e^{-2i(S,\lambda)} \right] \mathrm{SW}(\lambda). \qquad (9.35)$$

This is the famous Witten's 'magic formula' which expresses the Donaldson invariants in terms of Seiberg–Witten invariants. The Donaldson series then reads,

$$\mathbf{D}^{w_2(V)}(S) = 2^{2 + \frac{7\chi}{4} + \frac{11\sigma}{4}} \sum_\lambda e^{2i\pi(\lambda_0 \cdot \lambda + \lambda_0^2)} e^{S^2/2} e^{2(S,\lambda)} \mathrm{SW}(\lambda). \qquad (9.36)$$

Comparing to the structure theorem of Kronheimer and Mrowka (2.66) we find that the characteristic elements κ_λ are precisely the Seiberg–Witten basic classes 2λ, and the corresponding rational numbers a_λ are given by:

$$a_\lambda = 2^{2 + \frac{7\chi}{4} + \frac{11\sigma}{4}} e^{2i\pi(\lambda_0 \cdot \lambda + \lambda_0^2)} \mathrm{SW}(\lambda). \qquad (9.37)$$

These formulae agree with all known mathematical results. For example, for elliptic fibrations $E(n)$, $n \geq 2$, one can check (9.37) by comparing (2.70) and (3.28). From the above analysis we can deduce the following results for manifolds with $b_2^+ > 1$:

1) A simply connected four-manifold of Seiberg–Witten simple type is also of Donaldson simple type.

2) A non-simply connected manifold of Seiberg–Witten simple type is of strong simple type.

3) Any simply connected four-manifold with $b_2^+ > 1$ is of finite type, *i.e.*, (2.62) holds and the value of n depends on the maximal dimension of the Seiberg–Witten moduli spaces involved in the computation of (9.33).

In conclusion, the u-plane integral provides a physical way of computing the Donaldson invariants of four-manifolds of $b_2^+ > 0$, not necessarily of Seiberg–Witten simple type. In general it is the sum of two terms, the u-plane integral Z_u given for example in (8.68), and the Seiberg–Witten contributions at $u = \pm 1$ (which is essentially given by the expression (9.20) together with its transform (9.32)). When the manifold is of Seiberg–Witten simple type, the u-plane integral technology leads to a simple derivation of Witten's formula relating Donaldson and Seiberg–Witten invariants.

9.3. The blow-up formula

As we saw in Chapter 1, one basic operation that can be performed on a four-manifold in order to construct another four-manifold is the blow-up of a point. The behavior of Donaldson invariants under this operation has been investigated by many authors, and this effort culminated with the blow-up formulae of Fintushel and Stern. In this section we will obtain these blow-up formulae from the u-plane integral.

Remember that after blowing up a point in a four-manifold X we obtain another manifold $\widehat{X}$ with an extra two-homology class B with $B^2 = -1$ (we will denote by B its Poincaré dual as well). It then follows that $b_2^+(\widehat{X}) = b_2^+(X)$, $\chi(\widehat{X}) = \chi(X) + 1$, $\sigma(\widehat{X}) = \sigma(X) - 1$. Let $I(B)$ be the two-observable corresponding to B, and t a complex number. In order to compute the Donaldson invariants we have to choose as well an integral lifting of $w_2(V)$, $\widehat{\xi}$, in the blown up manifold. It is usually assumed that $\widehat{\xi}$ is a class that coincides with ξ away from B, and this means that $\widehat{\xi} = \xi + jB$ for $j = 0$ or 1. The goal of the blow-up formula is to compute

$$\langle \exp(2pu + I(S) + tI(B)) \rangle_{\widehat{\xi}, \widehat{X}}, \tag{9.38}$$

in a limit in which the area of B is small, in terms of

$$\langle \exp(2pu + I(S)) \rangle_{\xi, X}. \tag{9.39}$$

As pointed out by Moore and Witten in their work on the u-plane integral, it is natural to expect the existence of universal blow-up formulae based on physical considerations: when we blow up a point in X we produce an 'impurity' which is assumed to be very small (since in the blow-up formula the area of B is taken to be small), and we should be able to reproduce its effect by some local, $\overline{\mathcal{Q}}$-invariant observable. But in the twisted $\mathcal{N} = 2$ theory any local $\overline{\mathcal{Q}}$-invariant observable is a holomorphic function of u. Therefore there must be holomorphic functions $F_j(u, t)$, for $j = 0, 1$, such that

$$\langle \exp\left(2pu + I(S) + tI(B)\right)\rangle_{\widehat{\xi}, \widehat{X}} = \langle \exp\left(2pu + I(S) + F_j(u, t)\right)\rangle_{\xi, X}. \quad (9.40)$$

The above argument shows that the blow-up formula is universal, involving two holomorphic functions of u that do not depend on the four-manifold under consideration. This implies that one should be able to deduce the blow up formula from the u-plane integral alone. The reason for this is that one can find manifolds such that $Z_{\mathrm{SW}} = 0$ (where Z_{SW} denotes the contribution to Z_{DW} coming from $u = \pm 1$ and involving Seiberg–Witten invariants), and such that their blow-ups also have $Z_{\mathrm{SW}} = 0$. An example of those is $\mathbb{P}^2$ blown up at a small number of points: $\mathbb{P}^2$ admits a metric of positive curvature and has no chambers, so their Seiberg–Witten invariants vanish and $Z_{\mathrm{SW}} = 0$. The blown up $\mathbb{P}^2$ still admits a metric of positive curvature, and, moreover, if we blow it up at less than nine points, Seiberg–Witten wall crossing is still absent, since (9.12) is not satisfied. Therefore $Z_{\mathrm{SW}} = 0$ everywhere after blowing up, and for these manifolds the blow up functions must come from the u-plane integral.

The derivation of the blow-up formula from the u-plane integral is actually very easy, and can be obtained by comparing the u-plane integrand for X with that for $\widehat{X}$. Put $S \to S + tB$ and substitute into the integrand of (8.54), using as well the condition that we are in a chamber with $B_+ = 0$. The theta function (8.52) changes as follows:

$$\Psi_{\widehat{X}} = \Psi_X \exp\left[\frac{\pi t^2}{8\pi y}\left(\frac{du}{da}\right)^2\right] \sum_{n \in \mathbf{Z} + \frac{1}{2}w_2(V)\cdot B} \exp\left[i\pi\tau n^2 + int/h\right]e^{-i\pi n} \quad (9.41)$$

Similarly the measure factor (8.69) for the blown up manifold is related to that of the original manifold by:

$$\widehat{f}_{\widehat{X}} = \widehat{f}_X \vartheta_4^{-1} \exp\left[-t^2 \widehat{T}(u)\right]. \quad (9.42)$$

The ϑ_4^{-1} factor arises because the blow-up changes σ, and the other factor arises because $S^2 \to S^2 - t^2$. We then see that the blow-up has the effect of modifying the integrand by introducing an extra factor

$$\tau_0(t|u) = e^{-t^2 T(u)} \frac{\vartheta_4\left(\frac{t}{2\pi}\frac{du}{da}|\tau\right)}{\vartheta_4(0|\tau)} \tag{9.43}$$

in the case where $w_2(V) \cdot B = 0 \mod 2$ and a factor

$$\tau_1(t|u) = e^{-t^2 T(u)} \frac{\vartheta_1\left(\frac{t}{2\pi}\frac{du}{da}|\tau\right)}{\vartheta_4(0|\tau)} \tag{9.44}$$

when $w_2(V) \cdot B = 1 \mod 2$.

According to the physical argument above, these blow-up factors should depend only on u. We are going to prove that, indeed, they are series in t whose coefficients are polynomials in u.

Let us first consider the case $w_2(V) \cdot B = 0 \mod 2$. The quotient of theta functions can be written in terms of the Weierstrass sigma function:

$$\frac{\vartheta_4(z|\tau)}{\vartheta_4(0|\tau)} = e^{-2\eta_1\omega_1 z^2}\sigma_3(2\omega_1 z) \tag{9.45}$$

where ω_1 is proportional to the a-period, as shown in (7.72). Taking this into account, together with the identity

$$\eta_1 = \frac{\pi^2}{12\omega_1}E_2(\tau), \tag{9.46}$$

one finds that

$$\tau_0(t|u) = e^{-t^2 u/3}\sigma_3\left(\frac{4t}{\sqrt{2}}\right). \tag{9.47}$$

Now, the sigma functions $\sigma_i(t)$ can be expanded around the origin, and the coefficients of the Taylor expansion are polynomials in the roots e_i and the functions g_2, g_3 (in the Weierstrass description of the curve). For example,

$$\sigma_3(t) = 1 - e_3 t + \mathcal{O}(t^2), \tag{9.48}$$

As one can see in (7.69) and (7.71), these quantities only depend on u, proving our claim. Similarly, in the case $w_2(V) \cdot B = 1 \mod 2$ we use

$$\frac{\vartheta_1(z|\tau)}{\vartheta_1'(0|\tau)} = \frac{1}{2\omega_1}e^{-2\omega_1\eta_1 z^2}\sigma(2\omega_1 z) \tag{9.49}$$

and $\vartheta_1'(0|\tau) = \pi\vartheta_2(\tau)\vartheta_3(\tau)\vartheta_4(\tau)$, together with (7.103), to write the blow-up factor as

$$\tau_1(t|u) = \frac{\sqrt{2}}{4}e^{-t^2 u/3}\sigma\left(\frac{4t}{\sqrt{2}}\right). \tag{9.50}$$

The expansion of $\sigma(t)$ around the origin only involves g_2 and g_3,

$$\sigma(t) = t - \frac{g_2}{240}t^5 + \mathcal{O}(t^7), \tag{9.51}$$

therefore the blow-up factor (9.50) is also a function of u only.

A particularly interesting case of the blow-up formulae occurs at the degeneration points of the Seiberg–Witten curve, $u = \pm 1$. The blow up formula at those points describes the behavior of Donaldson invariants for manifolds of Donaldson simple type. Let us consider for simplicity the monopole point $u = 1$ (the behavior at $u = -1$ is identical). At $u = 1$ the period ω_1 diverges, and $\omega_3 = \pi\sqrt{2}i$. As shown in (B.2) the σ function degenerates to a trigonometric function, and one finds

$$\sigma(t) = e^{-t^2/6}\sinh t. \tag{9.52}$$

Similarly one finds that

$$\sigma_3(t) = e^{-t^2/6}\cosh t. \tag{9.53}$$

so the blow up functions become

$$\begin{aligned}\tau_0(t|u = 1) &= e^{-t^2/2}\cosh t, \\ \tau_1(t|u = 1) &= e^{-t^2/2}\sinh t,\end{aligned} \tag{9.54}$$

and they describe the behavior of Donaldson invariants under blow up for manifolds of Donaldson simple type. It can be seen that these formulae are compatible with Witten's formula (9.35): if $x = 2\lambda$ denotes the (integral) basic classes of the manifold of Seiberg–Witten simple type X, it can be shown that $\widehat{X}$ is also of Seiberg–Witten simple type. The basic classes of $\widehat{X}$ are given by $x \pm B$, and the Seiberg–Witten invariants satisfy $\mathrm{SW}(x \pm B) = \mathrm{SW}(x)$. It is immediate to see that (9.36) picks after blow-up the factors (9.54) depending on the values of $w_2(V) \cdot B$.

Bibliographical notes

- The applications of the u-plane integral developed in this chapter are owed to Moore and Witten [61]. The details of the extension to the non-simply connected case are given in [103]. On the mathematical side, the universal wall crossing formula in the simply connected case was obtained by L. Göttsche in [20]. In the non-simply connected case, explicit formulae have been obtained by V. Muñoz [21] for walls with $\lambda^2 = p_1/4$ and $\lambda^2 = p_1/4 + 1$.

- The phase depending on λ_0 in (9.20) is obtained from the duality transformation of the effective Lagrangian in [106].

- Witten's 'magic formula' relating Donaldson to Seiberg–Witten invariants was first written down in [22], and a complete derivation along physical lines was presented in [61]. In the Kähler case Witten was able in [107] to exploit the known behavior of $\mathcal{N} = 1$ super Yang–Mills theory to obtain a close ancestor of the magic formula. This ancestor involved the so called 'cosmic string contribution', which in the Kähler case is equivalent to the Seiberg–Witten invariants. Proving mathematically Witten's magic formula turns out to be a formidable task. A program to do that based on the non-abelian monopole equations which we analysed in Chapters 5 and 6 was proposed in [109], and developed (but so far not yet completed) by Feehan and Leness in a series of papers [64].

- The blow-up formulae derived in the text were first found in the context of Donaldson theory by Fintushel and Stern [110] (but notice that our normalizations differ from theirs). The behavior of Seiberg–Witten invariants under blow-up is described in [111].

Chapter 10

Further Developments in Donaldson–Witten Theory

In this final chapter we present some 'advanced' results in Donaldson–Witten theory, in considerably less detail than in the previous chapters. Our aim is to collect some useful results in the literature and provide an introduction to more specific developments. In the first section we present formulae for the Donaldson invariants of some four-manifolds with $b_2^+ = 1$ in special chambers. These formulae can be obtained from the u-plane integral by doing the integral on the fundamental domain. In the second section we consider the u-plane integral for theories with hypermultiplets, mostly in the case $b_2^+ > 1$, and explain some applications to the geography of four-manifolds. As in the rest of the book we focus on theories with gauge groups of rank one ($SU(2)$ or $SO(3)$). Finally, in the third section we consider the extension to gauge groups of rank larger than one.

10.1. More formulae for Donaldson invariants

In the previous chapter we have obtained three important properties of Donaldson invariants (the wall crossing formula, the structure of the Seiberg–Witten contribution, and the blow-up formula) without evaluating the integral over the fundamental domain. In this section we will discuss some results for Donaldson invariants that can be obtained by a more careful analysis of the u-plane integral.

A very important aspect of the u-plane integral is that in certain circumstances it can be shown to vanish. The precise statement is as follows: let X be a four-manifold with $b_2^+ = 1$ that fibers over a two-dimensional base B with a fiber Σ of genus zero. Then, if V is an $SO(3)$ bundle such that

$$w_2(V) \cdot [\Sigma] \neq 0, \tag{10.1}$$

then the u-plane integral vanishes when evaluated in the limiting chamber in which the area of the fiber is very small. As pointed out by Moore and Witten, this is easy to understand physically: consider for simplicity the case in which the four-manifold X is a direct product, $X = B \times \Sigma$. The Maxwell action which controls the $U(1)$ dynamics of the gauge field on the u-plane can then be written as

$$\int_{B \times \Sigma} F \wedge *F = \frac{\text{vol}\,(B)}{\text{vol}\,(\Sigma)} \left(\int_\Sigma F \right)^2 + \frac{\text{vol}\,(\Sigma)}{\text{vol}\,(B)} \left(\int_B F \right)^2, \qquad (10.2)$$

where F is the $U(1)$ field strength. The condition (10.1) means that $\int_\Sigma F \neq 0$, therefore as $\text{vol}\,(\Sigma) \to 0$ the Maxwell action diverges, causing the u-plane integral to vanish. However, one has to exercise caution, since although the *integrand* is clearly vanishing in these conditions, one has to worry about non-compactness of the moduli space. A careful analysis of the u-plane integral shows that, indeed, there is vanishing of the u-plane integral in these conditions. We refer the reader to the original work of Moore and Witten for a proof of this fact.

As an example of this vanishing result let us consider the product ruled surface $X_g = \mathbf{S}^2 \times \Sigma_g$, where Σ_g is a Riemann surface of genus g. $H^2(X_g, \mathbf{Z})$ is generated by two classes, $[\mathbf{S}^2]$ and $[\Sigma_g]$ (the Poincaré duals of the surfaces) and one has $[\mathbf{S}^2] \cdot [\Sigma_g] = 1$. The integral lift of $w_2(V)$ can be written as

$$w_2(V) = \epsilon[\Sigma_g] + \epsilon'[\mathbf{S}^2]. \qquad (10.3)$$

The period point ω of X_g was written in (2.76) in Chapter 2. The limiting chamber in which the volume of $\mathbf{S}^2$ is small corresponds to $\theta \to \infty$. It follows from the vanishing theorem that the u-plane integral vanishes in this limiting chamber provided that $w_2(V) \cdot [\mathbf{S}^2] \neq 0$, *i.e.*, for $\epsilon \neq 0$. Moreover, it is easy to see from the expression of the period point (2.76) that if we endow $\mathbf{S}^2$ and Σ_g with their standard constant curvature metrics the scalar curvature of X_g is positive for $e^{2\theta} > g - 1$. In particular, it will be positive in the limiting chamber of vanishing volume for $\mathbf{S}^2$, for any genus, therefore the Seiberg–Witten invariants will vanish in that chamber (owing to the vanishing theorem discussed in Chapter 3). We conclude that the Donaldson invariants of a product ruled surface vanish in the chamber of small volume for $\mathbf{S}^2$ for bundles with $w_2(V) \cdot [\mathbf{S}^2] \neq 0$.

The vanishing result is also useful in order to determine the Donaldson invariants of manifolds with $b_2^+ = 1$ in other chambers. The strategy is to consider first the Donaldson invariants in a chamber where they can be easily evaluated (for example, in a limiting chamber where one can show that the invariants vanish, as in the example of product ruled surfaces which we have considered). Then one can, in principle, sum the wall crossing terms needed to reach the chamber of interest. This procedure is useful as long as one can find an efficient procedure to sum up the wall crossing terms.

Another procedure for determining the Donaldson invariants is to evaluate the u-plane integral directly. It turns out that the integral over the fundamental domain which one ends up with in (8.54) is very similar to the integrals that compute one-loop corrections in string theory, and the same techniques that are used to compute these corrections can be applied to evaluate the u-plane integral. Again, we refer the reader to the original work of Moore and Witten, where these techniques are carefully developed in order to find explicit expressions for the u-plane integrals in certain chambers. Their results concern four-manifolds whose intersection contains as a summand the lattice (1.14).

We will now present results for a few examples where the techniques of Moore and Witten lead to explicit expressions for the Donaldson invariants. Consider again the product ruled surface X_g. Since its intersection form is precisely (1.14), the u-plane integral can be explicitly evaluated in the chamber where the volume of $\mathbf{S}^2$ is small (hence the Seiberg–Witten contribution vanishes), but for $w_2(V) \cdot [\mathbf{S}^2] = 0$.

Before presenting the result for the Donaldson–Witten generating series, let us clarify some issues related to the integrand of the u-plane integral (8.68) in the case of product ruled surfaces. As discussed in detail in Chapter 10, the integrand in the non-simply connected case involves a symplectic two-form Ω defined in (8.65). In the case of product ruled surfaces, this two-form is determined as follows: the basis of one forms on X_g is given by the duals to the usual symplectic basis of one cycles on Σ_g, δ_k, $k = 1, \ldots, 2g$, with $\delta_k \cap \delta_{k+g} = 1$, $k = 1, \ldots, g$. The matrix a_{kl} is then the symplectic matrix J, and Λ (the generator of the image of the map (8.64)) is given by $[\mathbf{S}^2]$. It follows that

$$\Omega = \sum_{k=1}^{g} \beta_k^{\sharp} \wedge \beta_{k+g}^{\sharp} . \tag{10.4}$$

and $\mathrm{vol}(\mathbf{T}^{b_1}) = 1$. It turns out that all the Donaldson polynomials involving the cohomology classes associated to one-cycles can be expressed in terms of the $Sp(2g, \mathbf{Z})$-invariant element in $\wedge^{\text{even}} H_1(X, \mathbf{Z})$ given by $\iota = -2 \sum_{k=1}^{g} \delta_k \delta_{k+g}$. This element of $\mathbf{A}(X)$ corresponds to the degree 6 differential form on the moduli space of instantons given by:

$$\gamma = -2 \sum_{i=1}^{g} I(\delta_i) I(\delta_{i+g}). \tag{10.5}$$

Writing $S = s\Sigma_g + t\mathbf{S}^2$, we present the results for $Z_{\text{DW}}(p, r, s, t) = \mathcal{D}_X^{w_2(E)}(e^{px+r\iota+s\Sigma_g+t\mathbf{S}^2})$. If we want to include ι in the u-plane integral, we just take into account that the 3-class $I(\delta_k)$ on the moduli space gives $(i/h)\beta_k^\sharp$ in the u-plane integral. Therefore using (10.4) we find the correspondence

$$\gamma \rightarrow \frac{2r}{h^2}\Omega, \tag{10.6}$$

and to obtain $Z_{\text{DW}}(p, r, s, t)$ from the above formulae we just have to change $(i/h)\delta^\sharp$ by (10.6) in (8.69).

We finally present the results for the Donaldson–Witten series of product ruled surfaces, in the limiting chamber of small volume for $\mathbf{S}^2$ and for $w_2(V) \cdot [\mathbf{S}^2] = 0$. There are two cases to consider. If $w_2(V) = 0$ the u-plane integral gives

$$Z_{\text{DW}}^{w_2(V)=0} = -\frac{i}{4}\left[(h^2 f_2)^{-1} e^{2pu+2stT(u)} \left(2f_1 h^2 s + 2r\right)^g \coth\left(\frac{is}{2h}\right) \right]_{q^0}, \tag{10.7}$$

where u, h, T, and $f_{1,2}$ are the modular forms introduced in Chapter 10 (all of them corresponding to the semi-classical cusp at infinity). If $w_2(E) = [\mathbf{S}^2]$ the u-plane integral gives

$$Z_{\text{DW}}^{w_2(V)=[\mathbf{S}^2]} = -\frac{1}{4}\left[(h^2 f_2)^{-1} e^{2pu+2stT(u)} \left(2f_1 h^2 s + 2r\right)^g \csc\left(\frac{is}{2h}\right) \right]_{q^0}. \tag{10.8}$$

For $g = 0$ one finds the results for the Donaldson–Witten series obtained by Göttsche and Zagier.

One of the main interest of considering Donaldson theory on product ruled surfaces is that one can perform a 'dimensional reduction' to a two-dimensional theory, and deduce properties of this low-dimensional model from the four-dimensional results. For example, it is easy to show that the moduli space of anti-self-dual connections on $\Sigma_g \times \mathbf{S}^2$ with instanton number zero is isomorphic to the moduli space of flat connections on Σ_g. The moduli space of flat $SO(3)$ connections on Σ_g with Stiefel–Whitney class $w_2 \neq 0$ is known to be a very rich and interesting space, and it can be realized in many different ways. Using the Hitchin–Kobayashi correspondence, for example, we can think about this space as the moduli space of rank two, odd degree stable bundles over Σ_g with fixed determinant. On the other hand, owing to the classical theorem of Narashiman and Seshadri we can identify this moduli space with the representations in $SU(2)$ of the fundamental group of the punctured Riemann surface $\Sigma_g \backslash D_p$, where D_p is a small disk around the puncture p, and with holonomy -1 around p (that we require a non-trivial holonomy is owed precisely to the non-zero Stiefel–Whitney class). In any case, this moduli space, which we will denote by $\mathcal{M}_g$, is a smooth projective variety of (real) dimension $6g - 6$.

The cohomology ring of $\mathcal{M}_g$ can be studied by using a two-dimensional version of the μ map which arises in Donaldson theory. This map sends homology classes of Σ_g to cohomology classes of $\mathcal{M}_g$. The generators of $H_*(\Sigma_g)$ give, in fact, a set of generators in $H^{4-*}(\mathcal{M}_g)$ which are usually taken as follows:

$$\alpha = 2\mu(\Sigma_g) \in H^2(\mathcal{M}_g),$$
$$\psi_i = \mu(\gamma_i) \in H^3(\mathcal{M}_g), \qquad (10.9)$$
$$\beta = -4\mu(x) \in H^4(\mathcal{M}_g),$$

where x is the class of the point in $H_0(\Sigma_g)$. We also define the $\mathrm{Sp}(2g, \mathbf{Z})$-invariant cohomology class in $H^6(\mathcal{M}_g)$,

$$\gamma = -2 \sum_{i=1}^{g} \psi_i \psi_{i+g}. \qquad (10.10)$$

In particular, the generators of the cohomology in (10.9) correspond precisely to the Donaldson cohomology classes, and we have that

$$\alpha = 2I(\Sigma_g), \qquad \psi_i = I(\gamma_i), \qquad \beta = -4\mathcal{O}, \qquad (10.11)$$

whilst the invariant form γ corresponds to (10.5). Our goal now is to evaluate the two-dimensional analogs of Donaldson invariants, $i.e.$, the intersection pairings

$$\langle \alpha^m \beta^n \gamma^p \rangle_{\mathcal{M}_g} = \int_{\mathcal{M}_g} \alpha^m \wedge \beta^n \wedge \gamma^p, \qquad (10.12)$$

as all the intersection pairings involving the ψ_is can be reduced to (10.12) by $\mathrm{Sp}(2g, \mathbf{Z})$ symmetry. Notice that this pairing is only different from zero when $2m + 4n + 6p = 6g - 6$. Since the Donaldson moduli space of X_g for zero instanton number agrees with $\mathcal{M}_g$, and the observables are also related as indicated in (10.11), the intersection pairings (10.12) and the Donaldson invariants of X_g are related as follows:

$$\langle \alpha^m \beta^n \gamma^p \rangle_{\mathcal{M}_g} = -\mathcal{D}^{w_2(V)=[\mathbf{S}^2]}_{\Sigma_g \times \mathbf{S}^2} \left((2\Sigma_g)^m (-4x)^n \iota^p \right). \tag{10.13}$$

The choice of Stiefel–Whitney class on the right hand side is made in order to induce $w_2 \neq 0$ on the bundle over the Riemann surface. The minus sign is because the left hand side used the orientation of the moduli space induced from its complex structure, and differs from the orientation inherited from Donaldson theory. Notice that the Donaldson invariants in (10.13) can be computed in any chamber, since for instantons with $c_2 = 0$ and $w_2(E) = [\mathbf{S}^2]$ one has $p_1 = 0$ and it follows from (9.7) that there are no walls. We will then use the explicit expression (10.8) to compute them. First of all, it follows immediately from (10.8) that the invariants of X_g are related to those of X_{g-1} through the relation

$$\frac{\partial}{\partial r} Z^{w_2(V)=[\mathbf{S}^2]}_{X_g} = 2g Z^{w_2(V)=[\mathbf{S}^2]}_{X_{g-1}}. \tag{10.14}$$

This implies the recursive relation

$$\langle \alpha^m \beta^n \gamma^\ell \rangle_{\mathcal{M}_g} = 2g \langle \alpha^m \beta^n \gamma^{\ell-1} \rangle_{\mathcal{M}_{g-1}}, \tag{10.15}$$

originally owed to Thaddeus. Since ℓ cannot be bigger than g in order to have a non-vanishing intersection pairing, we can use (10.15) to get rid of the γ operators. We now compute the intersection pairings $\langle \alpha^m \beta^n \rangle$. To do this we use the expansion:

$$\csc z = \sum_{k=0}^{\infty} (-1)^{k+1} (2^{2k} - 2) B_{2k} \frac{z^{2k-1}}{(2k)!}, \tag{10.16}$$

where B_{2k} are the Bernoulli numbers. We have to extract the powers $s^m p^n$ from the generating function (10.8). Since a power s^g comes already from the overall g-dependent factor in (10.8), we have to extract the power s^{m-g} from the series expansion in $s/2h$. Now taking into account the comparison factors from (10.13), and the dimensional constraint $2m + 4n = 6g - 6$, one finds

$$\langle \alpha^m \beta^n \rangle = \frac{1}{4} 2^m (-4)^n i^{m-g+1} 2^{2g+n-m} m! \frac{(2^{m-g+1} - 2)}{(m-g+1)!}$$
$$\times B_{m-g+1} [h_\infty^{3g-m-2} u_\infty^n f_{1\infty}^g f_{2\infty}^{-1}]_{q^0}. \tag{10.17}$$

Fortunately, only the leading term contributes in the q-expansion of this modular form, and one finally obtains,

$$\langle \alpha^m \beta^n \rangle = (-1)^g \frac{m!}{(m-g+1)!} 2^{2g-2} (2^{m-g+1} - 2) B_{m-g+1}, \qquad (10.18)$$

which is precisely Thaddeus' formula for the intersection pairings on $\mathcal{M}_g$.

So far we have considered product ruled surfaces in one limiting chamber in which $\mathbf{S}^2$ has small volume. The Donaldson invariants in the opposite limiting chamber in which Σ_g has small volume have been also computed for Stiefel–Whitney classes of the form $w_2(V) = [\mathbf{S}^2] + \epsilon[\Sigma_g]$, $\epsilon = 0, 1$, both by adding up an infinite series of wall crossing terms, and by computing directly the u-plane integral. For $g = 1$ one finds for example

$$Z_{\mathrm{DW}}^\epsilon(p, r, s, t) = -\frac{1}{2}(-1)^\epsilon \left[\frac{e^{2p+st}}{\cosh^2(t)} + (-1)^\epsilon \frac{e^{-2p-st}}{\cosh^2(-it)} \right]. \qquad (10.19)$$

This expression exhibits Donaldson simple type behavior, albeit in a very different way from the manifolds with $b_2^+ > 1$. The corresponding Donaldson series reads

$$\mathbf{D}^\epsilon = -(-1)^\epsilon \frac{e^{Q/2}}{\cosh^2 F}, \qquad (10.20)$$

where $F = \Sigma_1$. For $g > 1$ the Donaldson–Witten generating functional of X_g in the chamber of small volume for Σ_g and with $w_2(V) \cdot [\Sigma_g] \neq 0$ turns out to exhibit finite type behavior of order g, $i.e.$, it satisfies (2.62) with $n = g$. One reason to be particularly interested in the Donaldson invariants of product ruled surfaces in this chamber (and for $w_2(V) \cdot [\Sigma_g] \neq 0$) is because they are intimately related to the Gromov–Witten invariants of the moduli space of flat connections, $\mathcal{M}_g$. We refer the reader to the references at the end of this chapter for more details on this relationship.

We will now briefly consider our final example of a four-manifold with $b_2^+ = 1$, namely, $\mathbb{P}^2$. Since this manifold has $b_2 = 1$ there is no chamber structure. On the other hand, $\mathbb{P}^2$ admits a metric of positive scalar curvature (the usual Fubini–Study metric), so the Seiberg–Witten invariants vanish and the Donaldson invariants all come from the u-plane integral. The integration over the fundamental domain can be also explicitly performed in this case, although the techniques needed to do that are quite different from the ones used in the case of product ruled surfaces and one needs to use a non-holomorphic modular form constructed by Zagier. The result, owed to Moore and Witten, reads as follows. Define

$$\mathcal{H}(q) = \sum_{n\geq 0} H(4n)q^{2n} + \frac{1}{2}\sum_{n\geq 0} H(4n)q^{n/2}(-1)^n - \sum_{n>0} H(4n-1)q^{2n-1/2},$$
(10.21)

where $H(n)$ are Hurwitz class numbers, and define as well

$$\mathcal{H}(q,j) = \left(q\frac{d}{dq}\right)^j \mathcal{H}(q).$$
(10.22)

The Donaldson–Witten generating functional of $\mathbb{P}^2$ is then given by

$$Z_{\mathrm{DW}}(p,S) =$$

$$\mathrm{Res}_{q=0}\left(\frac{dq}{q}\cdot(-\frac{1}{2})\sum_{j=0}^{\infty}\frac{\Gamma(3/2)}{j!\Gamma(3/2+j)}\frac{S^{2j+1}}{2^j}\left[e^{2pu+S^2 T}\frac{\vartheta_4^9}{h^{4+2j}}\mathcal{H}(q,j)\right]\right).$$
(10.23)

It is interesting to notice that by comparing this explicit expression with another one obtained by Göttsche one obtains a new, explicit formula for $\mathcal{H}(q)$, therefore an explicit formula for Hurwitz class numbers. We refer the reader to the references at the end of this chapter for more details.

10.2. Applications to the geography of four-manifolds

In Chapter 5 we saw that one can consider extended versions of Donaldson–Witten theory by considering the coupling to twisted hypermultiplets, and that these models lead in general to non-abelian monopole equations like (5.64). When the gauge group is $G = SU(2)$ the underlying $\mathcal{N} = 2$ supersymmetric theories with $N_f \leq 4$ massive hypermultiplets in the fundamental representation have been analyzed by Seiberg and Witten following their work on pure $\mathcal{N} = 2$ supersymmetric Yang–Mills theory. For $N_f < 4$ these theories are still asymptotically free, while $N_f = 4$ is believed to correspond (for vanishing hypermultiplet masses) to a superconformal field theory. The moduli space of vacua of these theories has a Coulomb branch which can be parameterized by $u = \langle \mathrm{Tr}\,\phi^2\rangle$, so there is also a u-plane for the theories with gauge group $SU(2)$ and massive hypermultiplets. But for special values of the masses and when $N_f \geq 2$, there can also be Higgs branches in the moduli space of vacua, i.e., vacua where the scalar fields in the hypermultiplet take a vacuum expectation value different from zero. For *generic* values of the masses, the Higgs branch is absent and the picture for the moduli space is very similar to the case of pure Yang–Mills theory: there is a Coulomb branch or u-plane, and the low energy dynamics is described by an elliptic curve whose singularities correspond to particles that become massless and are located at the zeroes of the discriminant Δ_{N_f} of the curve. For generic masses there are in total $N_f + 2$ singularities.

Consider, for example, the case of one massive hypermultiplet, $N_f = 1$, whose mass will be denoted by m. According to Seiberg and Witten the low energy dynamics on the Coulomb branch is encoded in an effective action like (7.6), and the prepotential $\mathcal{F}(a, m)$ (which now depends on the mass m) can be obtained from the elliptic curve

$$y^2 = x^2(x - u) + \frac{1}{4}m\Lambda_1^3 x - \frac{1}{64}\Lambda_1^6, \tag{10.24}$$

where Λ_1 is the dynamically generated scale. The discriminant of the curve is given by

$$\Delta_1(u, m, \Lambda_1) = -u^3 + m^2 u^2 + \frac{9}{8}\Lambda_1^3 m u - \Lambda_1^3 m^3 - \frac{27}{256}\Lambda_1^6, \tag{10.25}$$

up to an overall numerical coefficient. We see that there are three singularities in the u plane, $u_j(m)$, $j = 1, 2, 3$, whose location depends on the value of the mass. As in the case of pure Yang–Mills, one defines a and $a_D = \partial\mathcal{F}/\partial a$ as periods of a certain meromorphic form λ_{SW} which is defined as well by (7.62). In the $N_f = 1$ case the explicit solution to this equation reads

$$\lambda_{\mathrm{SW}} = -\frac{\sqrt{2}}{8\pi}\frac{dx}{y}\left(3x - 2u + \frac{m\Lambda_1^4}{4x}\right), \tag{10.26}$$

and one can also use the technology developed in Chapter 7 to write a, a_D explicitly in terms of elliptic functions.

The observables of the twisted theory with hypermultiplets are the same as in Donaldson–Witten theory. We can then address the problem of computing the generating functional of correlators Z_{DW} defined in (5.48). It is easy to see that most of the results obtained in Chapter 8 can be transfered. Remember that in the pure Yang–Mills case the integrand of the u-plane integral was given by three pieces. The first piece involves the low energy effective Lagrangian and its form remains the same in the $SU(2)$ theories with hypermultiplets, the only difference being that the quantities involved in the Lagrangian are computed in the corresponding elliptic curve. For example, in the theory with $N_f = 1$ flavor da/du will be given again by (7.72), and the periods $\omega_{1,3}$ are computed for the elliptic curve (10.24).

The second piece involves the couplings to gravity, which give a measure of the form $A^\chi B^\sigma$. Since A is given by a period of the elliptic curve the argument of Chapter 8 will go through and we will still have (8.14). If we remember the derivation of the measure based on anomaly matching it is easy to see that B has to be proportional in general to the discriminant of the corresponding elliptic curve, since each singularity in the u-plane leads to a massless particle that has been integrated out and whose anomaly has to be reproduced by the measure. We then find that the general form of the measure for the theories with hypermultiplets is given by

$$A^\chi B^\sigma = \alpha_{N_f}^\chi \beta_{N_f}^\sigma \Delta_{N_f}^{\sigma/8} \left(\frac{du}{da}\right)^{\chi/2}, \qquad (10.27)$$

and α_{N_f}, β_{N_f} are in principle functions of the bare masses m_i of the hypermultiplets and of the dynamical scale Λ_{N_f}.

The third piece of the low energy effective action corresponds to the insertion of observables and their contact terms. Although the insertion of observables in (8.19) is given by the canonical solution to the descent equations, to find the contact terms we need additional arguments. In the case of theories with $N_f \leq 2$ hypermultiplets, it is easy to see that the expression for the contact term $T(u)$ turns out to be given again by (8.60). We conclude that the u-plane integral Z_u for $SU(2)$ theories with hypermultiplets is almost identical to the u-plane integral in the pure Yang–Mills case. An important difference is that in general the domain of integration is no longer given by a quotient of a subgroup of $\mathrm{Sl}(2, \mathbf{Z})$. All the properties that we derived from a knowledge of the integrand, such as blow-up and wall crossing formulae, can be generalized immediately to the theories with hypermultiplets, and we refer the reader to the work of Moore and Witten for details on this. The only real surprise which is found at this level is that for $N_f = 4$ there is no longer wall crossing behavior for manifolds of $b_2^+ = 1$, but, rather, continuous metric dependence.

The determination of the Seiberg–Witten contribution to the generating functional can be also done with the technique of cancellation of wall crossing that we explained in Chapter 8. This contribution is again a sum of terms associated to the different singularities of the curve, with each term having exactly the same form as in the pure Yang–Mills case given in (9.30). The functions C, L, P around the singularity at u_* are given by the obvious generalization of (9.24):

$$C = \frac{a - a_*}{q},$$

$$P = -\pi^2 \beta^8 \Delta_{N_f} (a - a_*)^{-1}, \qquad (10.28)$$

$$L = \pi i \alpha^4 \left(\frac{du}{da} \right),$$

where a is the appropriate local coordinate around u_*, and a_* is its value there (in such a way that $a - a_*$ vanishes at the singularity). For manifolds with $b_2^+ > 1$ the u-plane integral vanish, as in the pure Yang–Mills case, and the generating functional is given solely by the Seiberg–Witten contributions. For manifolds of simple type one can generalize the computation of the previous chapter to obtain an explicit formula for the generating function:

$$Z(m_i; p, S) = 2^{1 + \frac{3\sigma + \chi}{2}} (-i)^{\chi_h} \left(\frac{\pi^2 \beta_{N_f}^8}{2^8} \right)^{\sigma/8} \left(\frac{-\pi \alpha_{N_f}^4}{2} \right)^{\chi/4}$$

$$\times \sum_{j=1,\ldots,2+N_f} \kappa_j^{\chi_h} \left(\frac{da}{du} \right)_j^{-(\chi_h + \sigma)}$$

$$\times \sum_\lambda \mathrm{SW}(\lambda) \exp\left[2pu_j + S^2 T_j - i \left(\frac{du}{da} \right)_j (S, \lambda) \right] e^{2\pi i (\lambda_0^2 + \lambda \cdot \lambda_0)}.$$

$$(10.29)$$

For simplicity we have not included any insertion of $I_1(\delta)$. The sum over j is a sum over the $2 + N_f$ singularities at finite values on the u-plane. The sub-index j in the different quantities means that they are evaluated at the j-th singularity. As we explained in Chapter 5, the theories with matter can be considered on any smooth, compact, oriented four-manifold X if the non-abelian magnetic flux of the $SO(3)$ gauge bundle E satisfies $w_2(E) = w_2(X)$, where $w_2(E)$, $w_2(X)$ are the second Stiefel–Whitney classes of the bundle E and of the manifold X, respectively. Therefore in the expression (10.29) we have to choose an integer lifting $v = 2\lambda_0$ of $w_2(X)$. The quantity κ in (10.29) is given by

$$\kappa = \frac{du}{dq}. \qquad (10.30)$$

The evaluation of du/da at the singular points can be easily carried out by using (7.58). In order to evaluate κ at a singular point u_* notice that we can expand, in terms of the appropriate τ variable,

$$u = u_* + \kappa_* q + \mathcal{O}(q^2). \qquad (10.31)$$

By comparing now the q expansions of both sides of (7.100) we find

$$\kappa_* = \frac{(192\, g_2(u_*))^3}{\Delta'(u_*)}. \tag{10.32}$$

An interesting consequence of (10.29) is that, at least for generic masses and for manifolds of simple type with $b_2^+ > 1$, the topological information contained in the generating functional of twisted $SU(2)$ theories with hypermultiplets is the same as the one contained in topological $SU(2)$ Yang–Mills theory, and it is given by the Seiberg–Witten invariants. Of course, the generating functionals are different, since the quantities associated to the elliptic curves are different. We could say that the generating functionals of twisted theories with hypermultiplets contain a sort of universal part given by the Seiberg–Witten invariants, and a non-universal part which depends on the masses and number of flavors and can be computed from the corresponding Seiberg–Witten curves.

From the point of view of the topology of four-manifolds, the above result is somewhat disappointing since it implies that the information contained in the non-abelian $SU(2)$ monopole equations which describe the theory at high energies is identical to the information contained in the Seiberg–Witten invariants. However, it is natural to wonder if some new input coming from the physics of the theories with hypermultiplets can give new insights into the structure of the generating functional, therefore on the structure of Seiberg–Witten invariants. Indeed, this turns out to be the case, owing to the appearance of so called Argyres–Douglas points in these theories. Let us explain in some detail what is this new physical ingredient.

As we mentioned before, in the $SU(2)$ theories with N_f hypermultiplets and generic masses, there are $2 + N_f$ points in the moduli space where BPS states become massless (with one state becoming massless at each singularity). However, for special values of the masses these singularities can collide. The structure of the low energy theory at the point of collision depends in a crucial way on the charges of the colliding states. We will say that two BPS states with charges $\vec{n}_k = (n_e^k, n_m^k)$, $k = 1, 2$, are *mutually local* if $\vec{n}_1 \wedge \vec{n}_2 = 0$. It follows from this definition that if two states are mutually local one can always perform an $\text{Sl}(2, \mathbf{Z})$ duality rotation in such a way that the charges of both states are parallel, *i.e.*, there is a frame in which both particles are 'electric'. It follows from this definition that there are two possibilities for the collisions of singularities. The first possibility is that the k states which come together are mutually local, and the low energy theory will simply be $\mathcal{N} = 2$ supersymmetric abelian gauge theory with k hypermultiplets. The second possibility is that the colliding BPS states are not mutually local. The colliding singularities lead in this case to a non-trivial superconformal field theory in four dimensions, and the point in moduli space where this occurs is usually called a superconformal point or Argyres–Douglas point.

The simplest superconformal point occurs in the massive $N_f = 1$ theory. When $m = m_* = 3\Lambda_1/4$ and $u = u_* = 3\Lambda_1^2/4$, two singularities corresponding to a massless monopole and to a massless quark collide, giving rise to a superconformal point (there are two other superconformal points obtained by a $\mathbf{Z}_3$ rotation of this one). Since

$$g_2(u, m, \Lambda_1) = \frac{1}{4}\left(\frac{u^2}{3} - \frac{\Lambda_1^3 m}{4} \right),$$

$$g_3(u, m, \Lambda_1) = \frac{1}{16}\left(\frac{2u^3}{27} - \frac{\Lambda_1^3 mu}{12} + \frac{\Lambda_1^6}{64} \right) \tag{10.33}$$

this point corresponds to a complete degeneration of the elliptic curve where $g_2 = g_3 = 0$. If we write

$$m = m_* + z, \qquad u = u_* + \Lambda_1 z + \delta u, \tag{10.34}$$

and we introduce the shifted variable $x = u/3 + \tilde{x}$ the Seiberg–Witten curve (10.24) becomes, at leading order,

$$y^2 = \tilde{x}^3 - \frac{\Lambda_1^3}{4}z\tilde{x} - \frac{\Lambda_1^4}{16}\delta u, \tag{10.35}$$

which is a deformation of the cuspidal cubic $y^2 = \tilde{x}^3$. The variables z and δu correspond to operators in the superconformal field theory. The scaling dimensions of these operators can be deduced from (10.35) by dimensional analysis: since $a \sim (\delta u/y)d\tilde{x}$ has dimension 1 one finds that z has dimension $4/5$, while δu has dimension $6/5$. The various quantities associated with the elliptic curve, such as the periods and (10.30), will have a critical behavior as we approach the superconformal point, with critical exponents typical of this particular superconformal field theory (there are, in fact, other types of superconformal points in the theories with $N_f \geq 2$). Our purpose now is to analyse the generating functional of the twisted theory near the superconformal point, so we will start by analysing in detail the critical behavior of the quantities involved in (10.29).

Let us denote by $u_\pm = u_* + \Lambda_1 z + \delta u_\pm$ the position of the two colliding singularities. The deformation parameter $\delta u_\pm$ will depend on z as well, and of course $\delta u_\pm \to 0$ when $z \to 0$. The dependence of $\delta u_\pm$ on z can be obtained as a power series expansion by looking at the zeros of the discriminant (10.25), $\Delta_1(u, m, \Lambda_1) = 0$ in terms of the variables (10.34). The most appropriate normalization for Λ_1, for our purposes, is $4\sqrt{3}\Lambda_1^{3/2} = 1$ (with this normalization the leading term of $(du/da)^2$ is $z^{1/2}$.) One then finds

$$\delta u_\pm = \pm \left(\frac{16}{243}\right)^{1/3} z^{3/2} + \frac{4}{9}z^2 + \mathcal{O}(z^{5/2}). \tag{10.36}$$

Notice that

$$\delta u_-(z^{1/2}) = \delta u_+(-z^{1/2}). \tag{10.37}$$

This is because, for $z < 0$ one has $\pm z^{1/2} = \pm i|z|^{1/2}$. But in this case the roots $u_\pm$ of the equation $\Delta_1(m, u, \Lambda_1) = 0$ must be related by complex conjugation, therefore we must have (10.37). Using the expansion (10.36) and the explicit expressions (10.33), (7.58) and (10.32), we have the following expansions of $(du/da)^2$ and κ at the singularities $u_\pm$ for $m = m_* + z$:

$$\left(\frac{du}{da}\right)^2_\pm = \pm z^{1/2}\left(1 \pm \left(\frac{4}{3}\right)^{4/3} z^{1/2} + \mathcal{O}(z)\right)$$
$$\kappa_\pm = \mp 2^{31/3} \cdot 3^{4/3} z^{3/2}\left(1 \pm \left(\frac{2048}{3}\right)^{1/3} z^{1/2} + \mathcal{O}(z)\right). \tag{10.38}$$

Owing to (10.37) one has the following property:

$$\left(\frac{du}{da}\right)^2_-(z^{1/2}) = \left(\frac{du}{da}\right)^2_+(-z^{1/2}), \tag{10.39}$$

and a similar equation for $\kappa_\pm$. This will be important when we consider the analytic properties of the generating functional as a function of z. Notice that the leading powers of z in the expansion of δu, $(du/da)^2$ and κ are determined by the anomalous scaling weights of the operators near the superconformal point.

We can now focus on the analysis of the generating functional (10.29) for $N_f = 1$ near the superconformal point. First of all we have to make clear that as long as we are not at the superconformal point the contributions of each singularity to (10.29) are well defined and manifestly finite when we choose the appropriate local coordinate. Near the superconformal point, there are two colliding singularities with critical behavior, whilst the third singularity (where nothing special happens) will still give a finite answer. Therefore in order to analyse the critical behavior we can focus on the contributions of the colliding singularities. One can show that the critical behavior does not depend on the overall constants α_1, β_1, so in order to analyse the generating functional is enough to look at

$$F(z) = F_+(z) + F_-(z), \tag{10.40}$$

where

$$F_\pm(z) = \kappa_\pm^{\chi_h} \left[\left(\frac{du}{da} \right)_\pm^2 \right]^{\frac{\chi_h+\sigma}{2}} e^{2pu_\pm + S^2 T_\pm} \sum_\lambda \mathrm{SW}(\lambda) e^{-i(du/da)_\pm (S,\lambda)} e^{2\pi i(\lambda_0^2 + \lambda \cdot \lambda_0)}. \tag{10.41}$$

A priori it seems that $F_\pm(z)$ is a Laurent series in $z^{1/4}$, since $(du/da)_\pm$ has an expansion in $z^{1/4}$. However, a more detailed analysis of (10.41) shows that this can be refined. To see this consider the λ-dependent piece of $F_\pm(z)$:

$$\mathcal{SW}^\pm(z^{1/4}) = \sum_\lambda \mathrm{SW}(\lambda) e^{-i(du/da)_\pm (S,\lambda)} e^{2\pi i(\lambda_0^2 + \lambda \cdot \lambda_0)}. \tag{10.42}$$

An important property of these functions is that, if $\chi_h + \sigma$ is even (odd) they only contain even (odd) powers of $(du/da)_\pm$. This is because if λ is a basic class then $-\lambda$ is also a basic class and their Seiberg–Witten invariants are related by (3.26). On the other hand, changing λ to $-\lambda$ in the phase in (10.42) introduces a global factor

$$e^{-4\pi i \lambda_0 \cdot \lambda} = (-1)^\sigma, \tag{10.43}$$

which is a consequence of Wu's formula (1.22). If follows from this analysis that $F_\pm(z)$ only contains even powers of $(du/da)_\pm$, therefore (since χ_h is an integer), the functions $F_\pm$ have a series expansion in powers of $z^{1/2}$. Actually, more is true: owing to (10.39) we have $F_-(z^{1/2}) = F_+(-z^{1/2})$, therefore $F(z)$ has, in fact, an expansion in integral powers of z.

We can now study the properties of this function as a power series in z. It is clear that its behavior will depend on the topological properties of the four-manifold. First of all, by using (10.38) it is easy to see that the leading power of z in the factors which do not depend on λ in (10.41) is given by

$$\frac{c_1^2 - \chi_h}{4} = \frac{7\chi + 11\sigma}{16}. \tag{10.44}$$

Since (10.42) is regular at $z = 0$ and $F(z)$ has no monodromies it follows that $F(z)$ is regular at the origin if

$$\chi_h - c_1^2 - 4 < 0. \tag{10.45}$$

If this inequality is not satisfied the behavior at the origin will depend on (10.42) and, in particular, on the Seiberg–Witten invariants of the manifold. A simple analysis of (10.42) leads to the following *sufficient* condition for regularity of $F(z)$, which we will call the *superconformal simple type* (SST) condition: define the Seiberg–Witten series of X as the holomorphic function

$$\mathrm{SW}_X^v(\zeta) = \sum_x (-1)^{(v^2 + v \cdot x)/2} \mathrm{SW}(x) e^{x\zeta}, \tag{10.46}$$

where ζ is a formal variable, $x = 2\lambda$ are characteristic elements in $H^2(X, \mathbf{Z})$, and v is an integer lifting of $w_2(X)$. As in (2.67), the exponential e^x is understood here as a multilinear map on $\mathrm{Sym}^*(H^2(X, \mathbf{Z}))$. We will say that a manifold X of $b_2^+ > 1$ and of Seiberg–Witten simple type is of superconformal simple type (or, in short, that X is SST) if $\mathrm{SW}_X(\zeta)$ has a zero at $\zeta = 0$ of order $\geq \chi_h - c_1^2 - 3$. Roughly speaking, this zero compensates the pole associated with the λ independent prefactor in $F(z)$ in order to guarantee regularity of (10.41). Notice that a different choice of lifting v' will change (10.46) by a sign

$$(-1)^{\left((v' - v)/2\right)^2}, \tag{10.47}$$

so the SST property does not depend on the choice of lifting. Of course, any manifold that satisfies (10.45) is automatically SST.

The importance of the SST condition comes from the following: if a manifold X is SST then the generating functional (10.29) is finite. Now, on physical grounds we expect it, in fact, to be finite. The reason is the following: the only sources of divergences of correlation functions in a quantum field theory are the non-compactness of space-time or the non-compactness of moduli space. If X is compact, as we are assuming in our book, the first divergence does not occur, and since $b_2^+ > 1$ the contributions to the correlation functions come from a finite number of points in the u-plane. Moreover, for $N_f = 1$ another possible source of non-compactness (Higgs branches) are also absent. Therefore the generating functional should be finite. One can see that the contributions of the singularities $F_\pm(z)$ may diverge separately for manifolds which are SST (for example, they do diverge for a K3 surface), but when we add them together divergences cancel. Indeed, we are obliged to add both contributions owing to compactness of X. We then see that as soon as we put the twisted theory on a compact four-manifold, the singular behavior associated with the superconformal point disappears from physical quantities.

Although the SST condition is a sufficient, but not necessary, condition for analyticity of $F(z)$ around $z = 0$, it is natural to assume that four-manifolds avoid potential divergences of $F(z)$ at the origin by being SST. Therefore it is natural to conjecture that *every compact oriented four-manifold with $b_2^+ > 1$ of simple type is of superconformal simple type.* One can easily check that all minimal complex surfaces are SST. For example, the elliptic fibrations $E(n)$ with $n \geq 2$ have $b_2^+ > 1$ and are of Seiberg-Witten simple type, and they have $c_1^2 = 0$ and $\chi_h = n$, so that (10.45) is not satisfied for $n \geq 4$. However, the Seiberg-Witten series (10.46) can be easily computed by using (3.28) and choosing the lifting $v = c_1(K) = (n-2)[f]$, where K is the canonical line bundle of $E(n)$ and f is the class of the fiber. The result is simply

$$\mathrm{SW}_{E(n)}^{c_1(K)}(\zeta) = (2\sinh(\zeta f))^{\chi_h - 2}, \tag{10.48}$$

which indeed has a zero of order $\geq \chi_h - 3$. One can also check that all known geometric operations on four-manifolds (sucha as, for example, blowing up) preserve the SST condition. In fact, it has been proved by Feehan, Lenness, Kronheimer, and Mrowka that the above conjecture is true under some mild assumptions. We refer the reader to the bibliography at the end of this chapter for more information about the physical and mathematical aspects of the SST condition.

One interesting property of SST manifolds is that their number of Seiberg–Witten basic classes B (where we count λ and $-\lambda$ as a single basic class) is bounded from below. It is not difficult to show that if $B > 0$ then

$$B \geq \left[\frac{\chi_h - c_1^2}{2}\right], \tag{10.49}$$

where $[\cdot]$ is the integral part function. One has, in particular,

$$c_1^2 \geq \chi_h - 2B - 1. \tag{10.50}$$

The inequality (10.50) is called the *generalized Noether inequality*. The reason is its similarity to the Noether inequality for minimal surfaces of general

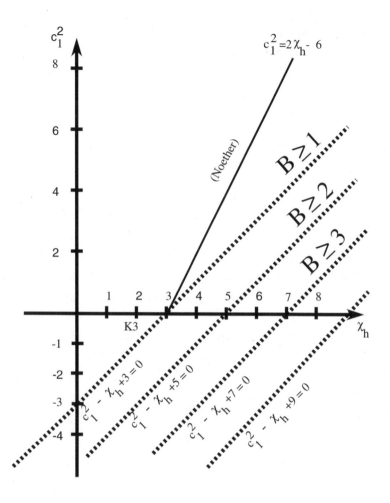

Fig. 1: Lines defining the generalized Noether inequalities.

type, which asserts that for these manifolds one has $c_1^2 \geq 2\chi_h - 6$. The above results, (10.49) and (10.50), make clear that for SST manifolds there are some interesting relationships between the classical topological invariants χ_h and c_1^2 and the Seiberg–Witten invariants. Indeed, one fundamental problem in the theory of four-manifolds is the so called *geography* problem: where do four-manifolds lie in the (χ_h, c_1^2) plane? The SST conjecture relates this problem to Seiberg–Witten theory. For example, the SST conjecture implies that four-manifolds of simple type and with $b_2^+ > 1$ lying below the line $c_1^2 - \chi_h + 3 = 0$ must have, either no basic classes (therefore a trivial generating functional), or at least two. This fact was observed experimentally by Fintushel and Stern, and as we have seen its rationale lies, surprisingly, in the physics of superconformal fixed points in $\mathcal{N} = 2$ supersymmetric theories (in fact, it is possible to show that if $F(z)$ is regular and X has one basic class, then *necessarily* $c_1^2 \geq \chi_h - 3$). The lines defining the generalized Noether inequality are shown in fig. 1 in the (χ_h, c_1^2) plane.

10.3. Extensions to higher rank gauge groups

In this book we have mostly discussed twisted $\mathcal{N} = 2$ supersymmetric gauge theories with gauge group $SU(2)$. The extension to gauge groups of higher rank is substantially more involved. From the mathematical point of view, this corresponds to the study of intersection theory on the moduli space of G-instantons, where G can, in principle, be any compact, semisimple Lie group. Although some of the results explained in Chapter 2 for $SU(2)$ extend to the general case (like the Atiyah–Hitchin–Singer deformation complex), this moduli space has never been studied in full detail from the point of view of intersection theory, and not much is known for Donaldson theory with gauge groups of rank larger than one. This additional complexity has a reflection in the physical approach which we have developed in this book. If the rank of the gauge group is $r > 1$ the integral over the Coulomb branch is no longer over a plane, but over $\mathbf{C}^r$, and it is much more difficult to provide explicit results for the Donaldson–Witten generating functional. There are however some interesting results for these theories that can be obtained using the physical approach, and in this last section we will summarize some of them.

The Lagrangian describing Donaldson–Witten theory with an arbitrary gauge group is written down in (5.28), and the topological algebra acting on the fields is written down in (5.29) and (5.32). As explained in Chapter 5, the observables of this model can be obtained from the descent procedure from any basis of gauge-invariant operators polynomial in ϕ. In the case of $SU(N)$ a convenient choice are the elementary symmetric polynomials U_k in the eigenvalues of ϕ, which are given by

$$U_k = \frac{1}{k}\text{Tr}\,\phi^k + \text{lower order terms}, \quad k = 2, \ldots, N. \tag{10.51}$$

From these operators one can form the operators $U_k^{(i)} = G^i U_k$ by using the canonical solution of the descent equations. For simplicity we will mainly focus on the observables associated to two-cycles,

$$I_k(S) = \int_S G^2 U_k. \tag{10.52}$$

Our goal is to compute in the physical approach the generating functional of Donaldson–Witten theory

$$Z(p_k, f_k, S) = \left\langle \exp\left(\sum_k (p_k U_k + f_k I_k(S)) \right) \right\rangle \tag{10.53}$$

where the vacuum expectation value is computed in the twisted theory, and the sum over k is over the labels of the gauge-invariant operators in ϕ.

In order to proceed with the physical approach to this model we need some properties of the low energy effective Lagrangian describing $SU(N)$ $\mathcal{N} = 2$ supersymmetric Yang–Mills theory with a rank r gauge group.

First of all, we describe the classical moduli space. As for $SU(2)$ this moduli space is given by the vacuum expectation values of the field ϕ, which can always be rotated into the Cartan subalgebra. If we denote by $\vec{\alpha}_I$ the simple roots of the Lie algebra, with $I = 1, \ldots, r$, these expectation values will be parameterized by a vector $\vec{a} = \sum_{I=1}^r a^I \vec{\alpha}_I$ in the root lattice. This theory also has BPS states, and their charges will be specified by vectors $\vec{q}$ expanded in the Dynkin basis (*i.e.*, the basis of fundamental weights). The central charges of electric BPS states are then written as

$$Z_{\vec{q}} = \vec{q} \cdot \vec{a}, \tag{10.54}$$

where the product is given by the usual bilinear form in the weight lattice. In order to obtain gauge-invariant coordinates on the moduli space, one can choose any basis of gauge-invariant operators made out of the field ϕ. The vacuum expectation values of these operators then provide the coordinates sought. For $SU(N)$ the usual choice is to take the correlation functions associated to (10.51)

$$u_k = \langle U_k \rangle. \tag{10.55}$$

As in the rank one case, the low energy effective action is determined by a prepotential $\mathcal{F}$ which depends now on r $\mathcal{N} = 2$ supersymmetric vector multiplets A^I. The vacuum expectation values of the scalar components of these vector superfields are the a^I. The dual variables and gauge couplings are defined as

$$a_{D,I} = \frac{\partial \mathcal{F}}{\partial a^I}, \qquad \tau_{IJ} = \frac{\partial^2 \mathcal{F}}{\partial a^I \partial a^J}. \tag{10.56}$$

The moduli space of vacua has a natural Kähler metric given by

$$(ds)^2 = \operatorname{Im} \tau_{IJ} da^I \, d\bar{a}^J, \tag{10.57}$$

which is invariant under the group $\mathrm{Sp}(2r, \mathbf{Z})$. This is the group of duality transformations of the low energy effective action, generalizing the $\mathrm{Sl}(2, \mathbf{Z})$ duality group of $SU(2)$. The symplectic group acts on the a^I, $a_{D,I}$ variables as $v \to \gamma v$, where $v^t = (a_{D,I}, a^I)$. If we write

$$\gamma = \begin{pmatrix} A & B \\ C & D \end{pmatrix}, \tag{10.58}$$

where A, B, C and D are $r \times r$ matrices, we have the following transformation properties,

$$\tau \to (A\tau + B)(C\tau + D)^{-1},$$
$$\operatorname{Im} \tau \to [(C\bar{\tau} + D)^{-1}]^t (\operatorname{Im} \tau)(C\tau + D)^{-1}. \tag{10.59}$$

$\mathcal{N} = 2$ supersymmetric Yang–Mills theory with gauge group G of rank r also contains monopoles and dyons in its spectrum. If we denote the charge of a dyon by $\vec{v} = (\vec{g}, \vec{q})$, where $\vec{g}$, $\vec{q}$ are the r-component vectors of magnetic and electric charges, respectively, then the central charge associated to it can be written in terms of the variables a^I, $a_{D,I}$ as

$$Z_{\vec{v}} = \vec{g} \cdot \vec{a}_D + \vec{q} \cdot \vec{a}, \tag{10.60}$$

generalizing the expression (10.54) for electric states.

The exact information about the low energy prepotential is encoded in an algebraic curve which generalizes the Seiberg–Witten curve for $SU(2)$. In the $SU(N)$ case this curve turns out to be hyper-elliptic and given by

$$y^2 = P(x)^2 - \Lambda^{2N}, \qquad P(x) = x^N - \sum_{I=2}^{N} u_I x^{N-I}, \tag{10.61}$$

which describes a Riemann surface of genus $N - 1$. For $N = 2$ this curve is of the form $y^2 = x^4 + \cdots$, but one can easily show that it is isogenous to the curve (7.61). Many concepts from the theory of elliptic curves extend to hyperelliptic curves. The polynomial in the right hand side of (10.61) has N roots e_i, and one can define the discriminant of the curve as $\prod_{i<j}(e_i - e_j)^2$. In the case of (10.61) the discriminant is given by

$$\Delta_\Lambda = \prod_{i<j} \Lambda^{2N^2} \Delta_0(u_2, \dots, u_{N-1}, u_N + \Lambda^N) \Delta_0(u_2, \dots, u_{N-1}, u_N - \Lambda^N), \tag{10.62}$$

where $\Delta_0 = \prod_{\vec{\alpha}_+} Z_{\vec{\alpha}_+}^2$, the product is over the positive roots, and $Z_{\vec{q}}$ is given in (10.54). As in the elliptic case, when two roots coincide and the discriminant has a zero the Riemann surface described by the hyper-elliptic curve becomes singular by pinching a one-cycle.

To fully specify the low energy effective action we also need a meromorphic differential defined on the hyper-elliptic curve, as in the rank one case. This differential λ_{SW}, also known as Seiberg–Witten differential, satisfies

$$\frac{\partial \lambda_{SW}}{\partial u^{I+1}} = \omega_I, \quad I = 1, \ldots, r, \tag{10.63}$$

where

$$\omega_I = \frac{x^{N-I-1} dx}{y} \tag{10.64}$$

is a basis of holomorphic differentials. λ_{SW} can be written explicitly as

$$\lambda_{SW} = \frac{1}{2\pi i} \frac{\partial P}{\partial x} \frac{x dx}{y}. \tag{10.65}$$

To compute the BPS masses a^I, $a_{D,I}$ one chooses a symplectic homology basis for the genus $N-1$ Riemann surface described by (10.61), $\alpha_I, \beta^I, I = 1, \ldots, r$. The generalization of (7.63) to the $SU(N)$ case is then

$$a_{D,I} = \oint_{\alpha_I} \lambda_{SW}, \quad a^I = \oint_{\beta^I} \lambda_{SW}. \tag{10.66}$$

If follows from this that

$$A_I{}^J = \oint_{\alpha_I} \omega_J = \frac{\partial a_{D,I}}{\partial u_{J+1}},$$
$$B^{IJ} = \oint_{\beta^I} \omega_J = \frac{\partial a^I}{\partial u_{J+1}}, \tag{10.67}$$

and the couplings τ_{IJ} are given by

$$\tau_{IJ} = A_I{}^K (B^{-1})_{KJ}. \tag{10.68}$$

Again as in the $SU(2)$ case, singularities in moduli space where the Riemann surface (10.61) degenerates are associated to massless dyons. The monodromy associated with a massless dyon of charge ν turns out to be given by

$$M_{\vec{\nu}} = \begin{pmatrix} 1 + \vec{q} \otimes \vec{g} & \vec{q} \otimes \vec{q} \\ -\vec{g} \otimes \vec{g} & 1 - \vec{g} \otimes \vec{q} \end{pmatrix}. \tag{10.69}$$

The structure of the singular loci in the Coulomb branch is much more complicated than in the $SU(2)$ case. There is a stratification structure, with submanifolds of codimension one where one BPS state becomes massless, and inside this submanifold there can be submanifolds of codimension 2 where two BPS become massless, and so on. In the $SU(N)$ case there are, in fact, N points where $N-1$ BPS states become massless. These states have zero electric charge and magnetic charges given by $\vec{g} = \vec{e}_i = (0, \ldots, 0, 1, 0, \ldots, 0)$, $i = 1, \ldots, N-1$, with the non-zero entry in the i-th position. At these points the low energy effective theory becomes a $U(1)^{N-1}$ theory with one massless monopole in each factor. These points are called the $\mathcal{N} = 1$ points since after breaking $\mathcal{N} = 2$ supersymmetry down to $\mathcal{N} = 1$ they give rise to the N vacua of $\mathcal{N} = 1$ super Yang–Mills theory. In the case of $SU(2)$, the $\mathcal{N} = 1$ vacua are precisely the points $u = \pm 1$. The moduli space of $SU(N)$ $\mathcal{N} = 2$ supersymmetric Yang–Mills theory also contains, for $N > 2$, points where at least two non-mutually local states become massless simultaneously. These are the Argyres–Douglas points which we discussed in the last section in the case of $SU(2)$ theories with massive hypermultiplets, and they give rise to superconformal field theories.

In any case, the principle 'Donaldson=Higgs+ Coulomb' also holds in the higher rank theories, therefore the generating functional (10.53) will have the structure

$$Z = Z_{\text{Coulomb}} + Z_{\mathcal{D}}. \tag{10.70}$$

The first piece is given by an integral over the Coulomb branch, and the other comes from the submanifolds $\mathcal{D}$ in which BPS states become massless and involves, in general, Seiberg–Witten invariants. One can show that the Coulomb branch contribution is different from zero only when $b_2^+(X) = 1$. The explicit expression of this integral can be worked out using the low energy effective action encoded in the prepotential, following steps very similar to those explained in Chapter 8, and reads

$$Z_{\text{Coulomb}} = \int_{\mathcal{M}_{\text{Coulomb}}} [da\,d\bar{a}] A(u_k)^\chi B(u_k)^\sigma e^{\sum p_k u_k + S^2 \sum f_k f_l T_{k,l}} \Psi. \tag{10.71}$$

The integrand of (10.71) has various ingredients. First of all, there is a gravitational part similar to (8.16). This part involves the factors:

$$A(u_k)^\chi = \alpha^\chi \left(\det \frac{\partial u_k}{\partial a^l} \right)^{\chi/2},$$

$$B(u_k)^\sigma = \beta^\sigma \Delta_\Lambda^{\sigma/8}. \tag{10.72}$$

In these equations α and β are constants. Notice that A and B in (10.72) are obtained from very natural quantities associated to the hyper-elliptic curve (10.61), namely, the determinant of the matrix of periods of ω_I, and the discriminant of the curve.

Another ingredient in (10.71) is Ψ, which is given by a sum over a lattice Γ. The lattice sum in Ψ comes essentially from the evaluation of the partition function of the photons in the effective $U(1)^r$ gauge theory. As in the $SU(2)$ case, we want to consider the possibility of turning on a non-abelian magnetic flux. A bundle E with gauge group $SU(N)$ is characterized up to isomorphism by two topological invariants: the instanton number and the generalized Stiefel–Whitney class (or non-abelian magnetic flux) $\vec{w}_2(E) \in H^2(X, \mathbf{Z}_N)$. For a general simply laced gauge group the magnetic fluxes are cohomology classes in $H^2(X, \Lambda_{\mathrm{w}}/\Lambda_{\mathrm{r}})$, where $\Lambda_{\mathrm{w(r)}}$ are the weight and root lattices of the group, respectively. For every weight lattice there is a set of weights called minimal weights which are in one-to-one correspondence with the cosets $\Lambda_{\mathrm{w}}/\Lambda_{\mathrm{r}}$. There are in general $c - 1$ minimal weights, where $c = \det C$ is the 'index of connection', that is, the determinant of the Cartan matrix (notice that c is precisely the order of Λ_{r} in Λ_{w}). The set of minimal weights is in general a subset of the set of fundamental weights. We will denote these weights by $\vec{m}_I$, $I = 1, \ldots, c - 1$. In the case of $SU(N)$ they are just the fundamental weights $\vec{w}_I$, $I = 1, \ldots, N - 1$. The electric line bundles are then classified by vectors of the form:

$$\vec{\lambda} = \vec{\lambda}_{\mathbf{Z}} + \vec{v}, \qquad \vec{\lambda}_{\mathbf{Z}} = \sum_{I=1}^{r} \lambda_{\mathbf{Z}}^I \vec{\alpha}_I, \qquad \vec{v} = \sum_{I=1}^{c-1} \pi^I \vec{m}_I, \tag{10.73}$$

where $\vec{\alpha}_I$ is a set of simple roots. In this expression $\lambda_{\mathbf{Z}}^I, \pi^I$ are all integer classes in $H^2(X; \mathbf{Z})$. The π^I are fixed and represent a choice of $\vec{w}_2(E) \in H^2(X, \Lambda_{\mathrm{w}}/\Lambda_{\mathrm{r}})$ lifted to $H^2(X, \Lambda_{\mathrm{r}})$. The set of elements of the form (10.73) is the lattice Γ involved in the Coulomb integral. Notice that we can always expand the minimal weights in the basis of simple roots:

$$\vec{m}_I = \sum_{J=1}^{r} m_I{}^J \vec{\alpha}_J, \quad I = 1, \ldots, c - 1, \quad m_I{}^J \in \frac{1}{c}\mathbf{Z}, \tag{10.74}$$

therefore we can write

$$\vec{\lambda} = \sum_{I=1}^{r} \lambda^I \vec{\alpha}_I, \quad \lambda^I = \lambda_{\mathbf{Z}}^I + \sum_{J=1}^{c-1} m_J{}^I \pi^J \in \frac{1}{c} H^2(X, \mathbf{Z}). \tag{10.75}$$

For $SU(N)$ one has $m_I{}^J = (C^{-1})_I{}^J$, where $C_I{}^J$ is the Cartan matrix. Finally, the generalization of (2.8) to the higher rank case is

$$c_2(E) = -\frac{\vec{v} \cdot \vec{v}}{2} \bmod 1. \tag{10.76}$$

In order to write Ψ let us introduce the following notation:

$$V_I \equiv \sum_k f_k \frac{\partial u_k}{\partial a^I}. \tag{10.77}$$

In terms of these quantities Ψ can be written as follows:

$$\Psi = (\det \mathrm{Im}\tau)^{-1/2} \exp\Big[\frac{1}{8\pi} V_J [(\mathrm{Im}\tau)^{-1}]^{JK} V_K S_+^2\Big]$$

$$\times \sum_{\vec{\lambda} \in \Gamma} \exp\Big[-i\pi\bar{\tau}_{IJ}(\lambda_+^I, \lambda_+^J) - i\pi\tau_{IJ}(\lambda_-^I, \lambda_-^J) - i\pi((\vec{\lambda} - \vec{\lambda}_0) \cdot \vec{\rho}, w_2(X))$$

$$- iV_I(S, \lambda_-^I)\Big]$$

$$\times \int \prod_{I=1}^r d\eta^I d\chi^I \exp\Big[-\frac{i\sqrt{2}}{16\pi} \mathcal{F}_{IJK} \eta^I \chi^J [4\pi(\lambda_+^K, \omega) + i(\mathrm{Im}\tau)^{KL} V_L(S, \omega)]$$

$$+ \frac{1}{64\pi} \overline{\mathcal{F}}_{KLI} (\mathrm{Im}\tau)^{IJ} \overline{\mathcal{F}}_{JPQ} \eta^K \chi^L \eta^P \chi^Q\Big]. \tag{10.78}$$

In this equation $\vec{\rho}$ is the Weyl vector (the sum of the fundamental weights), $w_2(X)$ is the second Stiefel–Whitney class of X, and η^I, χ^I are Grassmannian coordinates which arise from the zero-modes of the fermion fields in the theory. $\mathcal{F}_{IJK}$ denote the third derivatives of the prepotential. $\vec{\lambda}_0$ is an element in Γ such that $\vec{\lambda} - \vec{\lambda}_0 \in H^2(X, \Lambda_{\mathrm{r}})$, and, as in the $SU(2)$ case, it corresponds to a choice of orientation of the instanton moduli spaces. The phase factor $\exp[-i\pi((\vec{\lambda} - \vec{\lambda}_0) \cdot \vec{\rho}, w_2(X))]$ is the generalization of (8.15) to the higher rank case, and can be found, for example, from invariance of the Coulomb integral under the semi-classical monodromy. One should also include in the lattice sum a global phase factor depending on the generalized Stiefel–Whitney class $\vec{v}$ in order to obtain invariants which are real. We will discuss this factor later. Notice that in the rank one case (10.78) becomes precisely i multiplied by (8.53).

In (10.71) we have also included some terms of the form $T_{k,l}S^2$ in the exponential. These terms, which are proportional to the intersection form of the two-cycles, are the contact terms which generalize (8.60) in the higher rank case. In contrast to the rest of the terms in Ψ, the contact terms are not predicted *a priori* by the solution for the low energy effective action in terms of the hyper-elliptic curve and the Seiberg–Witten differentials. We will determine them later by using the behavior under blow-up, but for the time being let us understand the properties of the integrand under duality transformations. As we will see in a moment, this indeed provides some important information about the contact terms.

Let us introduce the following generalized Siegel–Narain theta function:

$$
\Theta_\Gamma(\tau_{IJ}, \alpha_I, \beta^I; P, \xi_I) =
$$

$$
\exp\left[-i\pi(\alpha_I, \beta^I) + \frac{\pi}{2}\left(\xi_{I,+}(\mathrm{Im}\tau)^{IJ}\xi_{J,+} - \xi_{I,-}(\mathrm{Im}\tau)^{IJ}\xi_{J,-}\right)\right]
$$

$$
\times \sum_{\bar{\lambda}\in\Gamma} \exp\left[-i\pi\bar{\tau}_{IJ}(\hat{\lambda}^I_+, \hat{\lambda}^J_+) - i\pi\tau_{IJ}(\hat{\lambda}^I_-, \hat{\lambda}^J_-) - 2\pi i(\hat{\lambda}^I, \xi_I) + 2\pi i(\hat{\lambda}^I, \alpha_I)\right],
$$

$$(10.79)$$

where $\hat{\lambda}^I = \lambda^I + \beta^I$. If we take

$$
\xi_I \equiv \frac{1}{2\pi}V_I S_- + \frac{\sqrt{2}}{16\pi}\overline{\mathcal{F}}_{IJK}\eta^J \chi^K \omega,
$$

$$
\beta^I = \sum_{I=1}^{c-1} m_I^J \pi^J, \qquad \alpha_I = \frac{1}{2}w_2(X), \quad I = 1,\ldots,r,
$$

$$(10.80)$$

and consider λ^I as the integer class $\lambda^I_{\mathbf{Z}}$ introduced in (10.73), the lattice sum Ψ can be written as

$$
\Psi = \exp\left[\frac{S^2}{8\pi}V_I[(\mathrm{Im}\,\tau)^{-1}]^{IJ}V_J\right](\det\mathrm{Im}\tau)^{-1/2}
$$

$$
\times \int \prod_{I=1}^{r} d\eta^I d\chi^I \exp\left[\frac{\sqrt{2}}{16\pi}\overline{\mathcal{F}}_{IJK}\eta^I\chi^J(\mathrm{Im}\,\tau)^{KL}V_L(S, \omega)\right] \qquad (10.81)
$$

$$
\Theta_\Gamma(\tau_{IJ}, \alpha_I, \beta^I; P, \xi_I).
$$

Define $\hat{\Psi} = \exp\left[-(S^2/8\pi)V_I[(\mathrm{Im}\,\tau)^{-1}]^{IJ}V_J\right]\Psi$. The Coulomb integral then reads:

$$
Z_{\mathrm{Coulomb}} = \int_{\mathcal{M}_{\mathrm{Coulomb}}} [du\,d\bar{u}]\exp\left[\left(\sum f_k f_l T_{k,l} + \frac{1}{8\pi}V_I[(\mathrm{Im}\,\tau)^{-1}]^{IJ}V_J\right)S^2\right]
$$

$$
\times \left|\det\left(\frac{\partial a^I}{\partial u_k}\right)\right|^2 A(u_k)^\chi B(u_k)^\sigma \hat{\Psi}.
$$

$$(10.82)$$

It can be checked that the factor in the second line of (10.82) is invariant under the symplectic duality group. We then see that this expression for the generating function (except for the exponential involving S^2) is the integral of a duality invariant object over a moduli space parametrized by the VEVs of the Casimirs, which are duality invariant coordinates. The only thing we need in order to achieve full invariance is invariance of the exponent involving S^2. This gives a constraint on the contact terms: the quantity

$$T_{k,l} + \frac{1}{8\pi}\frac{\partial u_k}{\partial a^I}[(\mathrm{Im}\tau)^{-1}]^{IJ}\frac{\partial u_l}{\partial a^J} \tag{10.83}$$

must be invariant under the action of the symplectic group $\mathrm{Sp}(2r, \mathbf{Z})$. Notice that (10.83) is the generalization of $\widehat{T}(u)$ in (8.69). As in the $SU(2)$ case, the other constraint on the contact terms has to do with their physical origin: these terms are quantum corrections and vanish at tree level, therefore they have to go to zero in the semi-classical region of moduli space (*i.e.*, when $\Lambda/a^I \to 0$). The duality transformation of the second term in (10.83) is easily worked out, and one finds that under a duality transformation it is shifted by:

$$-\frac{i}{4\pi}\frac{\partial u_k}{\partial a^I}\left[(C\tau + D)^{-1}C\right]^{IJ}\frac{\partial u_l}{\partial a^J}, \tag{10.84}$$

which again generalizes (8.57). The transformation of the contact term under an element of $\mathrm{Sp}(2r, \mathbf{Z})$ should compensate for this shift, therefore we have the transformation law

$$T_{k,l} \to T_{k,l} + \frac{i}{4\pi}\frac{\partial u_k}{\partial a^I}\left[(C\tau + D)^{-1}C\right]^{IJ}\frac{\partial u_l}{\partial a^J}. \tag{10.85}$$

The computation of the integral over the Coulomb branch is extremely involved, and so far it has not been attempted. However, we know from the analysis of the $SU(2)$ case that some important properties of the generating functional only involve the behavior of the integrand, and they are easier to analyse. One can, for example, analyse the wall crossing behavior in some detail, and we refer the reader to the bibliography at the end of this chapter for more information. Another aspect of the generating functional which only involves the integrand is the blow-up formula. We will analyse it here in the case of zero non-abelian magnetic flux. Suppose that we have a four-manifold X and we consider the blown up manifold at a point p, $\widehat{X}$. Since $\widehat{X}$ has an extra two-homology class, there are extra operators $I_k(B)$ that must be included in the generating function. We want to compute

$$\left\langle \exp\left[\sum_k (f_k I_k(S) + t_k I_k(B) + p_k \mathcal{O}_k)\right]\right\rangle_{\widehat{X}}, \qquad (10.86)$$

in terms of correlation functions of the twisted theory on X. In this equation we denoted $t_k \equiv f_k t$. The first thing to do is to analyse the change of the integrand under blow-up, and as in the $SU(2)$ case we assume that we are in a chamber where $(B, \omega) = 0$. The lattice sum changes then as follows:

$$\Psi_{\widehat{X}} = \left(\sum_{n^I} \exp\left[\pi i \tau_{IJ} n^I n^J + i \sum_k t_k \frac{\partial u_k}{\partial a^I} n^I - i\pi \sum_I n^I\right]\right)\Psi_X, \qquad (10.87)$$

and we see that

$$\Psi_{\widehat{X}} = \Theta[\vec{\Delta}, 0](\vec{\xi}|\tau)\Psi_X, \qquad (10.88)$$

where we have introduced the Riemann theta function $\Theta[\vec{\alpha}, \vec{\beta}](\vec{\xi}|\tau)$ with characteristics $\vec{\alpha} = (\alpha_1, \dots, \alpha_r)$, $\vec{\beta} = (\beta_1, \dots, \beta_r)$:

$$\Theta[\vec{\alpha}, \vec{\beta}](\vec{\xi}|\tau) = \sum_{n_K \in \mathbf{Z}} \exp\left[i\pi \tau_{IJ}(n_I + \beta_I)(n_J + \beta_J) + 2\pi i(n_I + \beta_I)(\xi_I + \alpha_I)\right]. \qquad (10.89)$$

The theta function $\Theta[\vec{\Delta}, 0](\vec{\xi}|\tau)$ involved in (10.88) has

$$\xi_I = \sum_k \frac{t_k}{2\pi} \frac{\partial u_k}{\partial a^I}, \quad \vec{\Delta} = (\frac{1}{2}, \dots, \frac{1}{2}). \qquad (10.90)$$

Notice that we have extracted from the Siegel–Narain theta function a standard theta function on the hyper-elliptic curve (10.61). The above expression is valid in the electric frame, and the characteristic of the theta function is inherited from the term $(\vec{\rho} \cdot \vec{\lambda}, w_2(X))$ in (10.78).

Let us now analyse the measure in the integrand. As $\chi(\widehat{X}) = \chi(X) + 1$ and $\sigma(\widehat{X}) = \sigma(X) - 1$, the measure picks an extra factor

$$\left(\det \frac{\partial u_J}{\partial a^I}\right)^{1/2} \Delta_\Lambda^{-1/8} = \frac{1}{\Theta[\vec{\Delta}, 0](0|\tau)}, \qquad (10.91)$$

as a consequence of Thomae's formulae for hyper-elliptic curves. Putting all these factors together we see that the blow-up factor in the integrand is given (up to a constant) by:

$$\tau(t_k|u_k) = e^{-\sum t_k t_l T_{k,l}} \frac{\Theta[\vec{\Delta}, 0](\vec{\xi}|\tau)}{\Theta[\vec{\Delta}, 0](0|\tau)}, \qquad (10.92)$$

which is the higher rank generalization of (9.43). Since (10.92) is an extra factor in the integrand of Z_{Coulomb} it follows by our arguments above that it must be invariant under duality transformations. Using the transformation law of the theta function under a symplectic transformation

$$\Theta[\vec{\alpha}, \vec{\beta}](\vec{\xi}|\tau) \to e^{i\phi}(\det(C\tau + D))^{1/2} \exp\left[\pi i \xi^t (C\tau + D)^{-1} C \xi\right] \Theta[\vec{\alpha}, \vec{\beta}](\vec{\xi}|\tau),$$
(10.93)

where $e^{i\phi}$ is a root of unity, one can easily prove that the duality invariance of the blow-up factor fixes the duality transformation of the contact terms, and one can re-derive (10.85) in this way.

The physical arguments which we gave in the third section of Chapter 9 indicate that the blow-up factor, in spite of its appearance, depends only on the parameters t_k and on the u_k, as we have indicated with our notation (and as we verified explicitly in the rank one case). More precisely, let $\vec{n} = (n_2, \ldots, n_N)$ be a vector of non-negative integers, with $|\vec{n}| = n_2 + \ldots + n_N$. There are then polynomials $\mathcal{B}_{\vec{n}}(u_2, \ldots, u_N)$ in the u_k such that:

$$\tau(t_k|u_k) = \sum_{|\vec{n}| \geq 0} t_2^{n_2} \cdots t_N^{n_N} \mathcal{B}_{\vec{n}}(u_2, \ldots, u_N)$$
(10.94)

and the following relation between the generating functions holds:

$$\left\langle \exp\left[\sum_k (f_k I_k(S) + t_k I_k(B) + p_k \mathcal{O}_k)\right]\right\rangle_{\tilde{X}} = \left\langle \exp\left[\sum_k (f_k I_k(S) + p_k \mathcal{O}_k)\right] \tau(t_k|\mathcal{O}_k)\right\rangle_X.$$
(10.95)

As a corollary of (10.94) we can derive an explicit expression for the contact terms by simply expanding (10.92) to second order in t_k. The first derivative of the theta function is zero owing to the choice of characteristic, and we find:

$$\tau(t_k|u_k) = 1 - \sum_{k,l} \left(T_{k,l} + \frac{1}{2\pi i}\partial_{\tau_{IJ}} \log \Theta[\vec{\Delta}, \vec{0}](0|\tau)\frac{\partial u_k}{\partial a^I}\frac{\partial u_l}{\partial a^J}\right) t_k t_l + \cdots,$$
(10.96)

Because of (10.94) this means that

$$T_{k,l} = -\frac{1}{2\pi i}\partial_{\tau_{IJ}} \log \Theta[\vec{\Delta}, \vec{0}](0|\tau)\frac{\partial u_k}{\partial a^I}\frac{\partial u_l}{\partial a^J} + \mathcal{B}_{\vec{n}_{k,l}}(u_2, \ldots, u_N),$$
(10.97)

where $\vec{n}_{k,l}$ are the vectors with $|\vec{n}_{k,l}| = 2$, corresponding to the quadratic terms in (10.94). The requirement that $T_{k,l}$ vanishes semi-classically implies that $\mathcal{B}_{\vec{n}_{k,l}}(u_2, \ldots, u_N) = 0$, and we finally obtain:

$$T_{k,l} = -\frac{1}{2\pi i}\partial_{\tau_{IJ}} \log \Theta[\vec{\Delta}, \vec{0}](0|\tau)\frac{\partial u_k}{\partial a^I}\frac{\partial u_l}{\partial a^J}, \tag{10.98}$$

which is the general expression for the contact terms owed to Losev, Nekrasov, and Shatashvili. The expansion (10.94) can, in fact, be worked out in detail and involves (not surprisingly in view of (9.47)) the hyper-elliptic generalization of sigma functions.

One can also generalize the arguments given in the $SU(2)$ case to derive the Seiberg–Witten contributions associated to BPS states. These are, in general, complicated, but when one considers the $\mathcal{N} = 1$ points in the $SU(N)$ theory (*i.e.*, the points in moduli space where $N - 1$ monopoles become massless) the Seiberg–Witten contribution can be written down quite explicitly. Let a^I be local coordinates for such a point, with $I = 1, \ldots, N-1$, and $q_{IJ} = \exp(2\pi i \tau_{IJ})$. The contribution of an $\mathcal{N} = 1$ point will be then given by a sum over Spin_c structures λ^I, $I = 1, \ldots, r$, with each r-tuple of λs contributing

$$e^{2\pi i(\lambda^I, \lambda_0^I)}\left(\prod_{I=1}^{r} \mathrm{SW}(\lambda^I)\right)$$

$$\times \mathrm{Res}_{a^1 = \cdots a^r = 0}\left[\left(\prod_{I=1}^{r}(a^I)^{(2\chi+3\sigma)/8-(\lambda^I)^2/2-1}\widetilde{q}_{II}^{\,-(\lambda^I)^2/2}\right)\right.$$

$$\times \prod_{1 \leq I < J \leq r}\left(q_{IJ}^{-(\lambda^I, \lambda^J)}\right)\left(\frac{\Delta_\Lambda}{\prod_{I=1}^{r} a^I}\right)^{\sigma/8}\left(\det \frac{\partial u_I}{\partial a^J}\right)^{\chi/2}$$

$$\left.\times \exp\left(\sum_k p_k u_k + S^2 \sum_{k,l} f_k f_l T_{kl} - i V_I(S, \lambda^I)\right)\right], \tag{10.99}$$

up to an overall constant (which depends on χ, σ). In (10.99) we have denoted $\widetilde{q}_{II} = q_{II}/a^I$. It is important to notice that the quantities $\widetilde{q}_{II}$ and q_{IJ} with $I \neq j$, as well as the factor involving $\Delta_\Lambda / \prod_{I=1}^{r} a^I$, are regular at $a^I = 0$.

If we now consider manifolds X with $b_2^+ > 1$, the only contribution to the generating functional comes precisely from the $\mathcal{N} = 1$ points. If, moreover, the Seiberg–Witten simple type condition holds one can evaluate the residue very easily by just evaluating the different quantities at the $\mathcal{N} = 1$ points.

This is done by using the explicit solution around the $\mathcal{N} = 1$ points owed to Douglas and Shenker together with the discrete $\mathbf{Z}_{4N}$ symmetry discussed in Chapter 7, which acts as $\mathbf{Z}_N$ in the Coulomb branch and relates the N $\mathcal{N} = 1$ vacua.

The $\mathcal{N} = 1$ points of $\mathcal{N} = 2$ supersymmetric theories are described by Chebyshev polynomials. At the point where $N - 1$ monopoles become massless, the polynomial $P_N(x)$ in (10.61) is of the form

$$P_N(x) = 2\cos\left(N \arccos \frac{x}{2}\right). \tag{10.100}$$

The branch points of the curve are now the single branch points $e_1 = -e_{2g+2} = 2$ and there are g double branch points

$$e_{2k} = e_{2k+1} = \widehat{\phi}_k = 2\cos\frac{\pi k}{N}. \tag{10.101}$$

The values of the u_k at this degeneration are given by the elementary symmetric polynomials in the eigenvalues

$$\phi_n = 2\cos\frac{\pi(n - \frac{1}{2})}{N}, \qquad n = 1, \ldots, N. \tag{10.102}$$

and one has for example,

$$u_2 = N, \qquad u_3 = 0, \qquad u_4 = \frac{N}{2}(3 - N), \tag{10.103}$$

and so on. Similarly, one can find the inverse period matrix

$$\frac{\partial u_{\ell+1}}{\partial a^I} = 2i(-1)^\ell \sin\frac{\pi I}{N} E_{\ell-1}(\widehat{\phi}_{p\neq m}), \tag{10.104}$$

where E_j is the elementary symmetric polynomial of degree j. Furthermore, one can find explicit expressions for the contact terms evaluated at this $\mathcal{N} = 1$ point, as well as for the off-diagonal gauge couplings

$$\tau_{IJ} = \frac{1}{\pi i} \log\frac{\gamma_J - \gamma_I}{\gamma_I + \gamma_J}, \quad I < J, \tag{10.105}$$

where

$$\gamma_I = \tan\frac{\pi I}{2N}. \tag{10.106}$$

We can already write the contribution from the $\mathcal{N} = 1$ points to the $SU(N)$ invariants, using the property that these points are related by the $\mathbf{Z}_{4N} \subset$

$U(1)_R$ symmetry. We must take into account the R-charges of the different operators in the correlation function, as well as of the measure in (10.99). We conclude that the generating functional in the case under consideration is given by

$$
Z(p_k, f_k, S) =
$$

$$
\sum_{k=0}^{N-1} \omega^{k[(N^2-1)\chi_h + N\vec{\lambda}_0 \cdot \vec{\lambda}_0]} \sum_{\lambda^I} e^{2\pi i (\lambda^I, \lambda_0^I)}
$$

$$
\times \exp\left[\sum_{\ell=2}^{N} p_\ell \omega^{k\ell} u_\ell - i\omega^{k(\ell-1)} \sum_{I=1}^{N-1} (S, \lambda^I) \frac{\partial u_\ell}{\partial a^I} + S^2 \sum_{\ell,\ell'} f_\ell f_{\ell'} \omega^{k(\ell+\ell'-2)} T_{\ell\ell'}\right]
$$

$$
\times \left(\prod_{I=1}^{N-1} \mathrm{SW}(\lambda^I)\right) \prod_{1 \le I < J \le N-1} \left(\frac{\gamma_I - \gamma_J}{\gamma_I + \gamma_J}\right)^{-2(\lambda^I, \lambda^J)},
$$

$$
\tag{10.107}
$$

up to some multiplicative constant $\tilde{\alpha}_N^\chi \tilde{\beta}_N^\sigma$, where $\tilde{\alpha}_N$ and $\tilde{\beta}_N$ are constants which depend on N. In (10.107), $\omega = \exp[i\pi/N]$, and we have included a phase factor $\omega^{kN\vec{\lambda}_0^2}$ depending on the generalized Stiefel–Whitney class, which generalizes the rank one case appearing in (9.32). This term can be obtained if we take into account that the instanton number of the bundle, once non-abelian fluxes are included, satisfies (10.76). The expression (10.107) can be seen to be real for any N if one multiplies it by the phase factor $e^{i\pi N\vec{v}^2/2}$, where $\vec{v}$ is the generalized second Stiefel–Whitney class.

The above formula generalizes Witten's formula (9.35) to the $SU(N)$ case. From a mathematical point of view (10.107) is somewhat disappointing, since it says that there is no new mathematical information in the $SU(N)$ generalization of Donaldson theory: the evaluation of the generating functional reduces again to the computation of Seiberg–Witten invariants, as in the case of $SU(2)$ theories with hypermultiplets of generic masses.

Bibliographical notes

- The vanishing theorem for the u-plane integral is derived in section 5 of [61]. The explicit computation of the u-plane integral uses techniques of string theory [112][113] generalized by Borcherds [114], and can be found in section 8 of [61]. The expressions presented in the text for Donaldson invariants of $\mathbf{S}^2 \times \mathbf{S}^2$ were obtained from Donaldson theory in [115] and from the u-plane integral in [61]. The extension to $\mathbf{S}^2 \times \Sigma_g$ was obtained, again from the u-plane integral, in [103] and [104]. Donaldson invariants of product ruled surfaces are also studied in [102][116].

- The applications of Donaldson theory to intersection theory on $\mathcal{M}_g$ explained in the text are based on [104], and they are also discussed in [102]. The relation between Donaldson theory on a product ruled surface and classical and quantum intersection theory on $\mathcal{M}_g$ is derived from a physical viewpoint in [117]. A good introduction to the topology and geometry of $\mathcal{M}_g$ can be found in [118]. The computation of intersection pairings on this moduli space is owed to Thaddeus [119]. The normalization of the generators of the cohomology ring in (10.9) follows [119] and [120]. A topological quantum field theory realization of the intersection theory on $\mathcal{M}_g$ can be found in [121] and [122], where the intersection ring is also computed with two-dimensional techniques.

- The quantum cohomology ring of $\mathcal{M}_g$ was predicted from physics in [117] and determined mathematically in [120]. A mathematical derivation of the Donaldson series of $\mathbf{S}^2 \times \mathbf{T}^2$ in the chamber of small volume for the two-torus can be found in [123][124].

- The Donaldson–Witten generating functional of $\mathbb{P}^2$ was obtained by Göttsche [20] by using the blow-up and wall crossing formulae. The expression in terms of class numbers presented in the text was obtained by Moore and Witten in [61] by using the u-plane integral, and by comparing with Göttsche's formula they were able to derive interesting new expressions for class numbers.

- An expression for the generating functional of the twisted $SU(2)$, $N_f = 1$ on Spin manifolds of simple type with $b_2^+ > 1$ was obtained in [125]. The general result was obtained by Moore and Witten in [61] by using the u-plane integral. Twisted theories with hypermultiplets are also considered in [102]. The $SU(2)$ theory with one massive hypermultiplet in the adjoint representation was analysed by using the u-plane in [126]. Twisted theories with massless hypermultiplets are considered in [99].

- Argyres–Douglas points were first found in the moduli space of $SU(3)$ $\mathcal{N} = 2$ supersymmetric Yang–Mills theories in [127], and later found to be present as well in the $SU(2)$ massive theories with hypermultiplets in [128]. The study of the massive $N_f = 1$ twisted theory around the Argyres–Douglas points, as well as the analysis of the consequences for the geography of four-manifolds, was performed in

[129], which also introduced the concept of manifolds of superconformal simple type. A summary of these results, intended for mathematicians, can be found in [130]. The SST conjecture is proved
up to some technical requirements in [131]. The geography of four-
manifolds is reviewed in [1]. The fact that manifolds with one basic
class seem to lie always above the line $c_1^2 - \chi_h + 3 = 0$ was observed
by Fintushel and Stern in [132], see [133] for more results along those
lines.

- The Seiberg–Witten solution was extended to the $SU(N)$ case in
  [134] and [135][136], and further explored in many other papers. The
  solution around the $\mathcal{N} = 1$ points is discussed in detail in [137].

- Donaldson–Witten theory with higher rank gauge group is discussed
  in [108] and [102], and reviewed in [100] and [138]. We mostly follow [108], which derives amongst other things the formula for the
  Donaldson–Witten generating functional in terms of Seiberg–Witten
  invariants presented in the text. A discussion of non-abelian magnetic fluxes can be found in [105]. The general form of the contact
  terms is obtained in [102]. There is an intimate relationship between
  the Seiberg–Witten solution and integrable hierarchies (see for example [139] and [140]) with implications for the twisted theories. A
  detailed analysis of the blow-up formula in the $SU(N)$ case, emphasizing the connection with integrability, is presented in [141] and
  reviewed in [142].

Appendix A

Spinors in Four Dimensions

In this appendix we collect the conventions used for spinors in both Minkowski and Euclidean spaces. In Minkowski space the flat metric has the form $\eta_{\mu\nu} = \text{diag}(-1,1,1,1)$, and the coordinates are labelled (x^0, x^1, x^2, x^3). The analytic continuation into Euclidean space is made through the replacement $x^0 = ix^4$ (and in momentum space, $p^0 = -ip^4$) the coordinates in this case being labelled (x^1, x^2, x^3, x^4) .

The Lorentz group in four dimensions, $SO(3,1)$, is not simply connected and therefore, strictly speaking, has no spinorial representations. To deal with these types of representations one must consider its double covering, the spin group $Spin(3,1)$, which is isomorphic to $SL(2,\mathbf{C})$. The group $SL(2,\mathbf{C})$ possesses a natural complex two-dimensional representation. Let us denote this representation by S and let us consider an element $\psi \in S$ with components $\psi_\alpha = (\psi_1, \psi_2)$ relative to some basis. The action of an element $M \in SL(2,\mathbf{C})$ is

$$(M\psi)_\alpha = M_\alpha{}^\beta \psi_\beta. \tag{A.1}$$

This is not the only action of $SL(2,\mathbf{C})$ which one could choose. Instead of M we could have used its complex conjugate $\overline{M}$, its inverse transpose $(M^{\text{T}})^{-1}$, or its inverse adjoint $(M^\dagger)^{-1}$. All of them satisfy the same group multiplication law. These choices would correspond to the complex conjugate representation $\overline{S}$, the dual representation $\widetilde{S}$, and the dual complex conjugate representation $\widetilde{\overline{S}}$. We will use the following conventions for elements of these representations:

$$\psi_\alpha \in S, \quad \psi_{\dot\alpha} \in \overline{S}, \quad \psi^\alpha \in \widetilde{S}, \quad \psi^{\dot\alpha} \in \widetilde{\overline{S}}. \tag{A.2}$$

These representations are called spinorial representations and are not independent. The $SL(2,\mathbf{C})$ invariant tensor $\epsilon_{\alpha\beta}$ (and $\epsilon_{\dot\alpha\dot\beta}$) allows one to relate a representation to its dual. We will use the following conventions for this tensor:

$$\epsilon_{21} = \epsilon^{12} = -\epsilon_{12} = -\epsilon^{21} = 1,$$
$$\epsilon_{\dot2\dot1} = \epsilon^{\dot1\dot2} = -\epsilon_{\dot1\dot2} = -\epsilon^{\dot2\dot1} = 1. \tag{A.3}$$

It allows one to raise and lower spinor indices:

$$\psi^\alpha = \epsilon^{\alpha\beta}\psi_\beta, \qquad \psi_\alpha = \epsilon_{\alpha\beta}\psi^\beta,$$
$$\psi^{\dot\alpha} = \epsilon^{\dot\alpha\dot\beta}\psi_{\dot\beta}, \qquad \psi_{\dot\alpha} = \epsilon_{\dot\alpha\dot\beta}\psi^{\dot\beta}. \tag{A.4}$$

One can easily verify that these relations are consistent with the assignment (A.2). Notice also that in our conventions $\epsilon_{\alpha\beta}\epsilon^{\beta\delta} = \delta^\beta_\alpha$, $\epsilon_{\dot\alpha\dot\beta}\epsilon^{\dot\beta\dot\delta} = \delta^{\dot\beta}_{\dot\alpha}$.

The real form of the complex Lie algebras $sl(2,\mathbf{C})$ and $su(2)_+ \times su(2)_-$ are the same. This allows one to use the notation (j_1, j_2) for the representations of $SL(2,\mathbf{C})$ where j_1 and j_2 are the spins of the two $su(2)$s. The representation S corresponds to $(\mathbf{2},\mathbf{0})$ whilst $\overline{S}$ to $(\mathbf{0},\mathbf{2})$. Notice that the two $su(2)$'s are not independent but related by complex conjugation. Under complex conjugation the labels of the representations must be interchanged. This allows one to restrict to representations which are fixed by complex conjugation in the case in which the two labels are the same. The resulting representation is real, and the simplest case, $(\mathbf{2},\mathbf{2})$, corresponds to the defining representation of the Lorentz group $SO(3,1)$ or vector representation.

The Clebsch–Gordan coefficients which intertwine between the vector representation and the $(\mathbf{2},\mathbf{2})$ of $SL(2,\mathbf{C})$ involve the Pauli matrices:

$$\sigma^1 = \begin{pmatrix} 0 & 1 \\ 1 & 0 \end{pmatrix}, \qquad \sigma^2 = \begin{pmatrix} 0 & -i \\ i & 0 \end{pmatrix}, \qquad \sigma^3 = \begin{pmatrix} 1 & 0 \\ 0 & -1 \end{pmatrix}, \tag{A.5}$$

and have the form

$$\sigma^\mu_{\alpha\dot\beta} = (1, \sigma^1, \sigma^2, \sigma^3)_{\alpha\dot\beta}, \tag{A.6}$$

allowing one to transform a four-vector $p_\mu = (p_0, p_1, p_2, p_3)$ into a bi-spinor:

$$\sigma^\mu p_\mu = \begin{pmatrix} p_0 + p_3 & p_1 - ip_2 \\ p_1 + ip_2 & p_0 - p_3 \end{pmatrix}. \tag{A.7}$$

The indices for the matrices σ^μ show how $SL(2,\mathbf{C})$ acts in bi-spinor space. If $M \in SL(2,\mathbf{C})$, the action of M is given by $\sigma^\mu p_\mu \to M(\sigma^\mu p_\mu)M^\dagger$, which preserves hermiticity and the determinant. The determinant is related to the Lorentz invariant quantity $p^\mu p_\mu = -\det(\sigma^\mu p_\mu) = -p_0^2 + p_1^2 + p_2^2 + p_3^2$ (the minus sign in front of the determinant is owed to our choice of metric). Notice that

M and $-M$ act in the same way on the vector representation, showing the fact that $SL(2, \mathbf{C})$ is a double cover of $SO(3, 1)$.

The dual Clebsch–Gordan coefficients are defined after raising the indices of σ^μ,

$$(\overline{\sigma}^\mu)^{\dot\alpha\beta} = \epsilon^{\dot\alpha\dot\beta}\epsilon^{\beta\alpha}\sigma^\mu_{\alpha\dot\beta} \qquad \longrightarrow \qquad (\overline{\sigma}^\mu) = (1, -\sigma^1, -\sigma^2, -\sigma^3) \qquad (A.8)$$

and allow to transform a vector into a bi-spinor corresponding to the dual representations. The matrices σ^μ and $\overline{\sigma}^\mu$ satisfy the following set of useful identities:

$$\begin{aligned}
\overline{\sigma}^\mu\sigma^\nu + \overline{\sigma}^\nu\sigma^\mu &= -2\eta^{\mu\nu}, \\
(\overline{\sigma}^\mu)^{\dot\alpha\beta}(\sigma_\mu)_{\gamma\dot\rho} &= -2\delta^\beta_\gamma\delta^{\dot\alpha}_{\dot\rho},
\end{aligned} \qquad (A.9)$$

which can be easily derived after using the following property of the Pauli matrices σ^i, $i = 1, 2, 3$:

$$\sigma^i\sigma^j = \delta^{ij} + i\epsilon^{ijk}\sigma^k \qquad (A.10)$$

where ϵ^{ijk} is the totally antisymmetric tensor with $\epsilon^{123} = 1$.

The generators of the Lorentz transformations can be constructed with the help of the matrices σ^μ and $\overline{\sigma}^\mu$. They take the form:

$$\begin{aligned}
(\sigma^{\mu\nu})_\alpha{}^\beta &= \frac{1}{4}\left(\sigma^\mu\overline{\sigma}^\nu - \sigma^\nu\overline{\sigma}^\mu\right)_\alpha{}^\beta, \\
(\overline{\sigma}^{\mu\nu})^{\dot\beta}{}_{\dot\alpha} &= \frac{1}{4}\left(\overline{\sigma}^\mu\sigma^\nu - \overline{\sigma}^\nu\sigma^\mu\right)^{\dot\beta}{}_{\dot\alpha},
\end{aligned} \qquad (A.11)$$

which leads to the simple form:

$$\begin{aligned}
\sigma^{0i} &= -\frac{1}{2}\sigma^i, & \sigma^{ij} &= -\frac{i}{2}\epsilon^{ijk}\sigma^k, \\
\overline{\sigma}^{0i} &= \frac{1}{2}\sigma^i, & \overline{\sigma}^{ij} &= -\frac{i}{2}\epsilon^{ijk}\sigma^k,
\end{aligned} \qquad (A.12)$$

and satisfy the properties:

$$\sigma^{\mu\nu} = \frac{i}{2}\epsilon^{\mu\nu\rho\sigma}\sigma_{\rho\sigma}, \qquad \overline{\sigma}^{\mu\nu} = -\frac{i}{2}\epsilon^{\mu\nu\rho\sigma}\overline{\sigma}_{\rho\sigma}, \qquad (A.13)$$

being $\epsilon_{0123} = 1$, $\epsilon^{0123} = -1$.

The matrices (A.11) provide spinorial representations of the Lorentz group:

$$M^{\mu\nu} \rightarrow i\sigma^{\mu\nu}, \qquad M^{\mu\nu} =\rightarrow i\overline{\sigma}^{\mu\nu}, \qquad (A.14)$$

satisfying the Lorentz algebra:

$$[M^{\mu\nu}, M^{\rho\tau}] = i\eta^{\mu\rho}M^{\nu\tau} - i\eta^{\mu\tau}M^{\nu\rho} - i\eta^{\nu\rho}M^{\mu\tau} + i\eta^{\nu\tau}M^{\mu\rho}. \qquad (A.15)$$

Under a Lorentz transformation parametrized by a real antisymmetric parameter $\omega_{\mu\nu}$ spinors transform as:

$$\delta\psi_\alpha = i\omega_{\mu\nu}(M^{\mu\nu})_\alpha{}^\beta\psi_\beta = -\omega_{\mu\nu}(\sigma^{\mu\nu})_\alpha{}^\beta\psi_\beta,$$
$$\delta\overline{\psi}_{\dot\alpha} = i\omega_{\mu\nu}(\overline{M}^{\mu\nu})_{\dot\alpha}{}^{\dot\beta}\overline{\psi}_{\dot\beta} = -\omega_{\mu\nu}(\overline{\sigma}^{\mu\nu})_{\dot\alpha}{}^{\dot\beta}\overline{\psi}_{\dot\beta}. \tag{A.16}$$

For the scalar product of spinors we will use the following conventions:

$$\psi\chi = \psi^\alpha\chi_\alpha = -\psi_\alpha\chi^\alpha = \chi\psi$$
$$\overline{\psi}\overline{\chi} = \overline{\psi}_{\dot\alpha}\overline{\chi}^{\dot\alpha} = -\overline{\psi}_{\dot\alpha}\overline{\chi}^{\dot\alpha} = \overline{\chi}\overline{\psi}. \tag{A.17}$$

In Euclidean space the relevant group is the orthonormal group $SO(4)$. In this case the covering group is $Spin(4)$, which is isomorphic to $SU(2)_+ \otimes SU(2)_-$. The representations are labelled by the spins of the two $SU(2)$s, which are independent. These spinorial representations are denoted by S_+ and S_-. The $SU(2)$ invariant tensor can be used to raise and lower indices following the same conventions (equations (A.3) and (A.4)) as in the Minkowskian case.

The Clebsh-Gordan coefficients which intertwine between the bi-spinorial representations and the defining representation of $SO(4)$, or vector representation, are

$$\sigma^\mu = (\sigma^1, \sigma^2, \sigma^3, i), \qquad \overline{\sigma}^\mu = (-\sigma^1, -\sigma^2, -\sigma^3, i). \tag{A.18}$$

Given a four vector (p_1, p_2, p_3, p_4) the negative of the determinant of its corresponding bi-spinor now becomes $-\det(\sigma^\mu p_\mu) = p_1^2 + p_2^2 + p_3^2 + p_4^2$, as expected. The matrices σ^μ and $\overline{\sigma}^\mu$ satisfy the identities

$$\overline{\sigma}^\mu\sigma^\nu + \overline{\sigma}^\nu\sigma^\mu = -2\delta^{\mu\nu},$$
$$(\overline{\sigma}^\mu)^{\dot\alpha\beta}(\sigma_\mu)_{\gamma\dot\rho} = -2\delta^\beta_\gamma\delta^{\dot\alpha}_{\dot\rho}. \tag{A.19}$$

In order to construct the generators of the rotation group $SO(4)$ one first defines the following matrices:

$$(\sigma^{\mu\nu})_\alpha{}^\beta = \frac{1}{4}(\sigma^\mu\overline{\sigma}^\nu - \sigma^\nu\overline{\sigma}^\mu)_\alpha{}^\beta,$$
$$(\overline{\sigma}^{\mu\nu})^{\dot\beta}{}_{\dot\alpha} = \frac{1}{4}(\overline{\sigma}^\mu\sigma^\nu - \overline{\sigma}^\nu\sigma^\mu)^{\dot\beta}{}_{\dot\alpha}, \tag{A.20}$$

One then easily finds out that the generators $M^{\mu\nu} = i\sigma^{\mu\nu}$ (or $M^{\mu\nu} = i\overline{\sigma}^{\mu\nu}$) satisfy the $SO(4)$ algebra:

$$[M^{\mu\nu}, M^{\rho\tau}] = i\delta^{\mu\rho}M^{\nu\tau} - i\delta^{\mu\tau}M^{\nu\rho} - i\delta^{\nu\rho}M^{\mu\tau} + i\delta^{\nu\tau}M^{\mu\rho}. \tag{A.21}$$

The explicit forms of the matrices $\sigma^{\mu\nu}$ and $\overline{\sigma}^{\mu\nu}$ are

$$\sigma^{\mu\nu} = \begin{pmatrix} 0 & -\frac{i}{2}\sigma^3 & \frac{i}{2}\sigma^2 & \frac{i}{2}\sigma^1 \\ \frac{i}{2}\sigma^3 & 0 & -\frac{i}{2}\sigma^1 & \frac{i}{2}\sigma^2 \\ -\frac{i}{2}\sigma^2 & \frac{i}{2}\sigma^1 & 0 & \frac{i}{2}\sigma^3 \\ -\frac{i}{2}\sigma^1 & -\frac{i}{2}\sigma^2 & -\frac{i}{2}\sigma^3 & 0 \end{pmatrix},$$

$$\tag{A.22}$$

$$\overline{\sigma}^{\mu\nu} = \begin{pmatrix} 0 & -\frac{i}{2}\sigma^3 & \frac{i}{2}\sigma^2 & -\frac{i}{2}\sigma^1 \\ \frac{i}{2}\sigma^3 & 0 & -\frac{i}{2}\sigma^1 & -\frac{i}{2}\sigma^2 \\ -\frac{i}{2}\sigma^2 & \frac{i}{2}\sigma^1 & 0 & -\frac{i}{2}\sigma^3 \\ \frac{i}{2}\sigma^1 & \frac{i}{2}\sigma^2 & \frac{i}{2}\sigma^3 & 0 \end{pmatrix}.$$

These matrices act as projectors of antisymmetric tensors and two-forms into their self-dual (SD) and anti-self-dual (ASD) parts. Given an antisymmetric tensor $V_{\mu\nu}$ one defines its SD part, $V_{\mu\nu}^+$, and ASD part, $V_{\mu\nu}^-$, as follows:

$$V_{\mu\nu}^{\pm} = \frac{1}{2}(V_{\mu\nu} \pm \frac{1}{2}\epsilon_{\mu\nu\rho\sigma}V^{\rho\sigma}), \tag{A.23}$$

where $\epsilon_{\mu\nu\rho\sigma}$ is the totally antisymmetric tensor corresponding to the volume form ($\epsilon_{1234} = 1$). The matrices $\sigma^{\mu\nu}$ and $\overline{\sigma}^{\mu\nu}$ satisfy the relations:

$$\sigma^{\mu\nu} = -\frac{1}{2}\epsilon_{\mu\nu\rho\sigma}\sigma^{\rho\sigma}, \qquad \overline{\sigma}^{\mu\nu} = \frac{1}{2}\epsilon_{\mu\nu\rho\sigma}\overline{\sigma}^{\rho\sigma} \tag{A.24}$$

and therefore they act as projectors into SD and ASD parts, providing the corresponding bi-spinor form:

$$\begin{aligned} V_{\alpha\beta} &= \epsilon_{\beta\gamma}(\sigma^{\mu\nu})_\alpha{}^\gamma V_{\mu\nu}^- = \epsilon_{\beta\gamma}(\sigma^{\mu\nu})_\alpha{}^\gamma V_{\mu\nu}, \\ V_{\dot{\alpha}\dot{\beta}} &= \epsilon_{\dot{\beta}\dot{\gamma}}(\overline{\sigma}^{\mu\nu})^{\dot{\gamma}}{}_{\dot{\alpha}} V_{\mu\nu}^+ = \epsilon_{\dot{\alpha}\dot{\gamma}}(\overline{\sigma}^{\mu\nu})^{\dot{\gamma}}{}_{\dot{\alpha}} V_{\mu\nu}. \end{aligned} \tag{A.25}$$

The tensors $V_{\alpha\beta}$ and $V_{\dot{\alpha}\dot{\beta}}$ are symmetric and transform as the $(\mathbf{3},\mathbf{0})$ and the $(\mathbf{0},\mathbf{3})$ representations of $SO(4)$. The antisymmetric tensor $V_{\mu\nu}$ can be also written in spinorial form as

$$V_{\alpha\dot{\alpha},\beta\dot{\beta}} = (\sigma^\mu)_{\alpha\dot{\alpha}}(\sigma^\nu)_{\beta\dot{\beta}}V_{\mu\nu}. \tag{A.26}$$

It turns out that the bi-spinor version of the decomposition of the antisymmetric tensor into SD and ASD parts takes the form:

$$V_{\mu\nu} = V_{\mu\nu}^+ + V_{\mu\nu}^- \quad \longrightarrow \quad V_{\alpha\dot{\alpha},\beta\dot{\beta}} = \epsilon_{\dot{\alpha}\dot{\beta}}V_{\alpha\beta} + \epsilon_{\alpha\beta}V_{\dot{\alpha}\dot{\beta}}. \tag{A.27}$$

For Euclidean spinors we will take $M_{\dot{\alpha}} = (M_{\dot{1}}, M_{\dot{2}})$ to be a positive chirality spinor. The complex conjugate spinor is defined as

$$\overline{M}^{\dot{\alpha}} = (\overline{M}^{\dot{1}}, \overline{M}^{\dot{2}}) = (M_{\dot{1}}^*, M_{\dot{2}}^*). \tag{A.28}$$

Appendix B

Elliptic Functions and Modular Forms

In this Appendix we recall the definitions and properties of some of the elliptic functions and modular forms used in the book, and we list some useful formulae.

Given a lattice with half-periods ω and ω' one can define the following elliptic functions:

$$
\sigma(z) = z \prod{}' \left(1 - \frac{z}{s}\right) e^{z/s + z^2/2s^2},
$$

$$
\zeta(z) = \frac{1}{z} + \sum{}' \left(\frac{1}{z-s} + \frac{1}{s} + \frac{z}{s^2}\right) = \frac{d}{dz} \log \sigma(z), \qquad \text{(B.1)}
$$

$$
\wp(z) = \frac{1}{z^2} + \sum{}' \left(\frac{1}{(z-s)^2} - \frac{1}{s^2}\right) = -\zeta'(z),
$$

where $s = 2m\omega + 2m'\omega'$, m, m' are integers, and $'$ indicates that the product or sum is over all pairs of integers (m, m') except $(0,0)$. The above functions are called, respectively, sigma, zeta, and Weierstrass functions. In the limit in which one of the periods goes to infinity, say ω', they become elementary functions:

$$
\sigma(z) = \frac{2\omega}{\pi} e^{\frac{1}{6}\left(\frac{\pi z}{2\omega}\right)^2} \sin \frac{\pi z}{2\omega},
$$

$$
\zeta(z) = \frac{1}{3}\left(\frac{\pi}{2\omega}\right)^2 z + \frac{\pi}{2\omega} \cot \frac{\pi z}{2\omega}, \qquad \text{(B.2)}
$$

$$
\wp(z) = -\frac{1}{3}\left(\frac{\pi}{2\omega}\right)^2 + \left(\frac{\pi}{2\omega}\right)^2 \frac{1}{\sin^2 \frac{\pi z}{2\omega}}.
$$

The incomplete elliptic integrals of the first and the second kind are, respectively,

$$
F(\phi, k) = \int_0^{\sin \phi} \frac{1}{\sqrt{(1-t^2)(1-k^2 t^2)}},
$$

$$
E(\phi, k) = \int_0^{\sin \phi} \sqrt{\frac{1 - k^2 t^2}{1 - t^2}}. \qquad \text{(B.3)}
$$

The complete elliptic integrals are defined by:

$$K(k) = F(\pi/2, k), \quad E(k) = E(\pi/2, k). \tag{B.4}$$

The Eisenstein function of weight two E_2 is defined by

$$E_2 = 1 - 24 \sum_{n=1}^{\infty} \frac{nq^n}{1-q^n} = 1 - 24q + \cdots \tag{B.5}$$

and transforms under $\mathrm{Sl}(2, \mathbf{Z})$ as follows:

$$E_2(\tau) \to (c\tau + d)^2 \left(E_2(\tau) + \frac{12c}{2\pi i(c\tau + d)} \right). \tag{B.6}$$

Our conventions for the Jacobi theta functions are:

$$
\begin{aligned}
\vartheta_1(\nu|\tau) &= i \sum_{n \in \mathbf{Z}} (-1)^n q^{\frac{1}{2}(n+1/2)^2} e^{i\pi(2n+1)\nu}, \\
\vartheta_2(\nu|\tau) &= \sum_{n \in \mathbf{Z}} q^{\frac{1}{2}(n+1/2)^2} e^{i\pi(2n+1)\nu}, \\
\vartheta_3(\nu|\tau) &= \sum_{n \in \mathbf{Z}} q^{\frac{1}{2}n^2} e^{i\pi 2n\nu}, \\
\vartheta_4(\nu|\tau) &= \sum_{n \in \mathbf{Z}} (-1)^n q^{\frac{1}{2}n^2} e^{i\pi 2n\nu},
\end{aligned}
\tag{B.7}
$$

where $q = e^{2\pi i\tau}$. When $\nu = 0$ we will simply denote $\vartheta_2(\tau) = \vartheta_2(0|\tau)$ (notice that $\vartheta_1(0|\tau) = 0$). The theta functions $\vartheta_2(\tau)$, $\vartheta_3(\tau)$ and $\vartheta_4(\tau)$ have the following product representation:

$$
\begin{aligned}
\vartheta_2 &= 2q^{1/8} \prod_{n=1}^{\infty} (1 - q^n)(1 + q^n)^2, \\
\vartheta_3 &= \prod_{n=1}^{\infty} (1 - q^n)(1 + q^{n-1/2})^2, \\
\vartheta_4 &= \prod_{n=1}^{\infty} (1 - q^n)(1 - q^{n-1/2})^2
\end{aligned}
\tag{B.8}
$$

and they have the following properties under modular transformations:

$$
\begin{aligned}
\vartheta_2(-1/\tau) &= \sqrt{\frac{\tau}{i}} \vartheta_4(\tau), & \vartheta_2(\tau + 1) &= e^{i\pi/4} \vartheta_2(\tau), \\
\vartheta_3(-1/\tau) &= \sqrt{\frac{\tau}{i}} \vartheta_3(\tau), & \vartheta_3(\tau + 1) &= \vartheta_4(\tau), \\
& & \vartheta_4(\tau + 1) &= \vartheta_3(\tau). \\
\vartheta_4(-1/\tau) &= \sqrt{\frac{\tau}{i}} \vartheta_2(\tau),
\end{aligned}
\tag{B.9}
$$

We also have the following useful identities:

$$\vartheta_3^4(\tau) = \vartheta_2^4(\tau) + \vartheta_4^4(\tau), \tag{B.10}$$

and

$$\vartheta_2^4(\tau)\vartheta_3^4(\tau)\vartheta_4^4(\tau) = 16\,\eta^{12}(\tau), \tag{B.11}$$

where

$$\eta(\tau) = q^{1/24} \prod_{n=1}^{\infty} (1 - q^n) \tag{B.12}$$

is the Dedekind eta function.

The expressions of the quantities involved in the Seiberg–Witten solution are, in terms of modular forms:

$$
\begin{aligned}
u &= \frac{1}{2} \frac{\vartheta_2^4 + \vartheta_3^4}{(\vartheta_2 \vartheta_3)^2}, \\
u^2 - 1 &= \frac{1}{4} \frac{\vartheta_4^8}{(\vartheta_2 \vartheta_3)^4} = \frac{\vartheta_4^8}{64 h^4(\tau)}, \\
\frac{du}{d\tau} &= \frac{\pi}{4i} \frac{\vartheta_4^8}{(\vartheta_2 \vartheta_3)^2}, \\
h(\tau) &= \frac{da}{du} = \frac{1}{2}\vartheta_2 \vartheta_3.
\end{aligned}
\tag{B.13}
$$

The first few terms in the q-expansions are:

$$
\begin{aligned}
u &= \frac{1}{8q^{1/4}} \left(1 + 20q^{1/2} - 62q + 216q^{3/2} + \cdots \right) \\
&= \frac{1}{8\,q^{1/4}} + \frac{5\,q^{1/4}}{2} - \frac{31\,q^{3/4}}{4} + 27\,q^{5/4} - \frac{641\,q^{7/4}}{8} + \frac{409\,q^{9/4}}{2} + \cdots,
\end{aligned}
\tag{B.14}
$$

$$u_M(q_D) = 1 + 32\,q_D + 256\,q_D^2 + 1408\,q_D^3 + 6144\,q_D^4 + 22976\,q_D^5 + 76800\,q_D^6 + \cdots, \tag{B.15}$$

$$
\begin{aligned}
T(u) &= -\frac{1}{24}\left[\frac{E_2}{h(\tau)^2} - 8u \right] \\
&= q^{1/4} - 2\,q^{3/4} + 6\,q^{5/4} - 16\,q^{7/4} + 37\,q^{9/4} - 78\,q^{11/4} \\
&\quad + 158\,q^{13/4} - 312\,q^{15/4} + 594\,q^{17/4} + \cdots, \\
T_M(q_D) &= \frac{1}{2} + 8\,q_D + 48\,q_D^2 + 224\,q_D^3 + 864\,q_D^4 + 2928\,q_D^5 + 9024\,q_D^6 + \cdots,
\end{aligned}
\tag{B.16}
$$

$$h = \frac{1}{2}\vartheta_2\vartheta_3 = \frac{1}{4}\vartheta_2^2(\tau/2)$$

$$= q^{1/8} + 2\,q^{5/8} + q^{9/8} + 2\,q^{13/8} + 2\,q^{17/8} + 3\,q^{25/8} + 2\,q^{29/8} + \cdots,$$

$$h_M = \frac{1}{2i}\vartheta_3\vartheta_4 = \frac{1}{2i}\vartheta_4^2(2\tau_D)$$

$$= \frac{1}{2i}(1 - 4q_D + 4q_D^2 + 4q_D^4 - 8q_D^5 + \cdots),$$

$$\text{(B.17)}$$

For a_D we also have the following expansion:

$$a_D(q_D) = 16iq_D(1 + 6q_D + 24q_D^2 + 76q_D^3 + \cdots), \qquad \text{(B.18)}$$

Bibliography

[1] R. Gompf and A. Stipsicz. (1999). *Four-manifolds and Kirby calculus*, (American Mathematical Society).

[2] S.K. Donaldson and P.B. Kronheimer. (1990). *The geometry of four-manifolds*, (Oxford Mathematical Monographs).

[3] A. Beauville. (1996). *Complex algebraic surfaces*, 2nd edition, (Cambridge University Press).

[4] W. Barth, C. Peters and A. van den Ven. (1984). *Compact complex surfaces*, (Springer-Verlag).

[5] R. Fintushel and R.J. Stern. (1998). 'Constructions of smooth four-manifolds', *Doc. Math.*, extra volume ICM, p. 443, math.GT/9907178.

[6] D. Freed and K. Uhlenbeck. (1991). *Instantons and four-manifolds*, (Springer-Verlag), second edition.

[7] R. Friedman and J.W. Morgan. (1994). *Smooth four-manifolds and complex surfaces*, (Springer-Verlag).

[8] R. Friedman and J.W. Morgan, editors. (1998) *Gauge theory and the topology of four-manifolds*, IAS/Park City Mathematics Series, (American Mathematical Society).

[9] S. Cordes, G. Moore and S. Rangoolam. (1996). 'Lectures on 2D Yang–Mills theory, equivariant cohomology and topological field theories', in *Fluctuating geometries in statistical mechanics and field theory*, (eds. F. David, P. Ginsparg and J. Zinn-Justin), Les Houches Session LXII, (Elsevier) p. 505, hep-th/9411210.

[10] G. 't Hooft. (1979). 'A property of electric and magnetic flux in non-abelian gauge theories', *Nucl. Phys.*, **B153**, 141.

[11] M.F. Atiyah, N.H. Hitchin and I.M. Singer. (1978). 'Self-duality in four dimensional Riemannian geometry', *Proc. Roy. Soc.*, **A362**, 425.

[12] P.B. Kronheimer and T.S. Mrowka. (1994). 'Recurrence relations and asymptotics for four-manifold invariants', *Bull. Amer. Math. Soc.*, **30**, 215; (1995). 'Embedded surfaces and the structure of Donaldson's polynomial invariants', *J. Diff. Geom.*, **41**, 573.

[13] R. Fintushel and R. Stern. (1995). 'Donaldson invariants of 4-manifolds with simple type', *J. Diff. Geom.*, **42**, 577.

[14] R. Stern. (1998). 'Computing Donaldson invariants', in *Gauge theory and the topology of four-manifolds*, IAS/Park City Math. Ser., (American Mathematical Society).

[15] V. Muñoz. (1999). 'Fukaya-Floer homology of $\Sigma_g \times \mathbf{S}^1$ and applications', *J. Diff. Geom.*, **53**, 279, math.DG/9804081.

[16] P.B. Kronheimer and T.S. Mrowka. (1997). 'The structure of Donaldson's invariants for four-manifolds not of simple type', unpublished.

[17] V. Muñoz. (2000). 'Basic classes for four-manifolds not of simple type', *Commun. Anal. Geom.*, **8**, (2000) 653, math.DG/9811089.

[18] V. Muñoz. (2002). 'Donaldson invariants of non-simple type four-manifolds', *Topology*, **41**, 745, math.DG/9909165.

[19] D. Kotschick and J. Morgan. (1994). '$SO(3)$-invariants for 4-manifolds with $b_2^+ = 1$, part II', *J. Diff. Geom.*, **39**, 433.

[20] L. Göttsche. (1996). 'Modular forms and Donaldson invariants for four-manifolds with $b_+ = 1$', *J. Am. Math. Soc.*, **9**, 827, alg-geom/9506018.

[21] V. Muñoz. (2000). 'Wall-crossing formulae for algebraic surfaces with positive irregularity', *J. Lond. Math. Soc.*, **61** 259, alg-geom/9709002.

[22] E. Witten. (1994). 'Monopoles and four-manifolds', *Math. Res. Lett.*, **1**, 769, hep-th/9411102.

[23] N. Seiberg and E. Witten. (1994). 'Electric-magnetic duality, monopole condensation, and confinement in $\mathcal{N} = 2$ supersymmetric Yang–Mills theory', *Nucl. Phys.*, **B426**, 19, hep-th/9407087.

[24] N. Seiberg and E. Witten. (1994). 'Monopole condensation, duality and chiral symmetry breaking in $\mathcal{N} = 2$ supersymmetric QCD', *Nucl. Phys.*, **B431**, 484, hep-th/9408099.

[25] J. W. Morgan. (1996). *The Seiberg-Witten equations and applications to the topology of smooth four-manifolds*, (Princeton University Press).

[26] M. Marcolli. (1999). *Seiberg-Witten gauge theory*, (Hindustan Book Agency).

[27] L. I. Nicolaescu. (2000). *Notes on Seiberg-Witten theory*, (American Mathematical Society).

[28] S. Donaldson. (1996). 'The Seiberg-Witten equations and four-manifold topology', *Bull. Amer. Math. Soc.*, **33**, 45.

[29] R. Dijkgraaf. (1996). 'Lectures on four-manifolds and topological gauge theories', *Nucl. Phys. Proc. Suppl.*, **B45**, 29.

[30] G. Thompson. (1992). 'New results in topological field theory and abelian

gauge theory', in *Trieste HEP Cosmology 1995*, p. 194, hep-th/9511038

[31] J.M.F. Labastida and C. Lozano. (1998). 'Lectures in topological quantum field theory', (eds. H. Falomir, R. Gamboa and F. Schaposnik), Proceedings of the CERN-Santiago de Compostela-La Plata Meeting on *Trends in Theoretical Physics*, (American Institute of Physics, New York), CP 419, 54-94; hep-th/9709192.

[32] O. García-Prada. (1998). 'Seiberg-Witten invariants and vortex equations', in *Quantum symmetries*. (eds. A. Connes, K. Gawedzki and J. Zinn-Justin), Les Houches Session LXIV, (North-Holland) p. 885.

[33] C. Okonek and A. Teleman. (1996). 'Seiberg-Witten invariants for manifolds with $b_2^+ = 1$, and the universal wall-crossing formula', *Int. J. Math.*, **7**, 811, alg-geom/9603003.

[34] R. Friedman and J.W. Morgan. (1997). 'Algebraic surfaces and Seiberg-Witten invariants', *J. Alg. Geom.*, **6**, (1997) 445, alg-geom/9502026; (1999). 'Obstruction bundles, semiregularity, and Seiberg-Witten invariants', *Commun. Anal. Geom.*, **7**, 451, alg-geom/9509007.

[35] R. Brussee. (1996). 'The canonical class and the C^∞ properties of Kähler surfaces', *New York J. Math.*, **2**, 103, alg-geom/9503004.

[36] O. Biquard. (1998). 'Les équations de Seiberg-Witten sur une surface complexe non Kählérienne', *Commun. Anal. Geom.*, **6**, 173.

[37] C. H. Taubes. (1994). 'The Seiberg-Witten invariants and symplectic forms', *Math. Res. Lett.*, **1**, 809; (1996). 'SW ⇒ Gr: From the Seiberg-Witten equations to pseudo-holomorphic curves', *J. Amer. Math. Soc.*, **9**, 845.

[38] P.B. Kronheimer and T.S. Mrowka. (1994). 'The genus of embedded surfaces in the projective plane," *Math. Res. Lett.*, **1**, 797.

[39] T.J. Li and A. Liu. (1995). 'General wall-crossing formula', *Math. Res. Lett.*, **2**, 797.

[40] J. Wess and J. Bagger. (1992). *Supersymmetry and supergravity*, Second edition, (Princeton University Press).

[41] S. J. Gates, M. T. Grisaru, M. Rocek and W. Siegel. (1983). *Superspace or One Thousand and One Lessons on Supersymmetry*, (The Benjamin/Cummings Pubishing Company, Inc., London).

[42] L. Álvarez-Gaumé and S.F. Hassan. (1997). 'Introduction to S-duality in $\mathcal{N} = 2$ supersymmetric gauge theories', *Fortsch. Phys.*, **45**, 159, hep-th/9701069.

[43] E. Witten. (1988). 'Topological quantum field theory', *Comm. Math. Phys.*, **117**, 353.

[44] D. Birmingham, M. Blau, M. Rakowski and G. Thompson. (1991). 'Topological field theory', *Phys. Rep.*, **209**, 129.

[45] A. Schwarz. (1978). 'The partition function of degenerate quadratic functional and Ray–Singer invariants', *Lett. Math. Phys.*, **2**, 247.

[46] E. Witten. (1989). 'Quantum field theory and the Jones polynomial', *Commun. Math. Phys.*, **121**, 351.

[47] E. Witten. (1988). 'Topological sigma models', *Comm. Math. Phys.*, **118**, 411.

[48] E. Witten. (1982). 'Constraints on supersymmetry breaking', *Nucl. Phys.*, **B202**, 253.

[49] E. Witten. (1982). 'Supersymmetry and Morse theory', *J. Diff. Geom.*, **17**, 661.

[50] L. Baulieu and I. Singer. (1988). 'Topological Yang–Mills symmetry', *Nucl. Phys. Proc. Suppl.*, **5B**, 12.

[51] H. Kanno. (1989). 'Weil algebra structure and geometrical meaning of BRST transformations in topological quantum field theory', *Z. Phys.*, **C43**, 477.

[52] J.P. Yamron. (1988). 'Topological actions from twisted supersymmetric theories', *Phys. Lett.*, **B213**, 325.

[53] D. Anselmi and P. Frè. (1993). 'Topological twist in four dimensions, R-duality and hyperinstantons', *Nucl. Phys.*, **B404**, 288, hep-th/921121; (1994). 'Topological sigma models in four dimensions and triholomorphic maps', *Nucl. Phys.*, **B416**, 255, hep-th/9306080; (1995). 'Gauged hyperinstantons and monopole equations', *Phys. Lett.*, **B347**, 247, hep-th/9411205.

[54] M. Álvarez and J.M.F. Labastida. (1993). 'Breaking of topological symmetry', *Phys. Lett.*, **B 315**, 251, hep-th/9305028; (1995). 'Topological matter in four dimensions', *Nucl. Phys.*, **B 437**, 356, hep-th/9404115.

[55] J.M.F. Labastida and M. Mariño. (1995). 'A topological Lagrangian for monopoles on four-manifolds', *Phys. Lett.*, **B351**, 146, hep-th/9503105.

[56] J.M.F. Labastida and M. Mariño. (1995). 'Non-abelian monopoles on four-manifolds', *Nucl. Phys.*, **B448**, 373, hep-th/9504010.

[57] S. Hyun, J. Park and J.S. Park. (1995). 'Topological QCD', *Nucl. Phys.*, **B453**, 199, hep-th/9503201.

[58] M. Mariño. (1997). 'The geometry of supersymmetric gauge theories in four dimensions', Ph. D. Thesis, hep-th/9701128.

[59] C. Lozano. (1997). *Teorías supersimétricas y teorías topológicas,* Tesina, Universidade de Santiago de Compostela.

[60] J.M.F. Labastida and M. Mariño. (1997). 'Twisted baryon number in $\mathcal{N} = 2$ supersymmetric QCD', *Phys. Lett.*, **B400**, 323, hep-th/9702054.

[61] G. Moore and E. Witten. (1998). 'Integrating over the u-plane in Donaldson theory', *Adv. Theor. Math. Phys.*, **1**, 298; hep-th/9709193.

[62] C. Okonek and A. Teleman. (1995). 'The coupled Seiberg-Witten equations, vortices, and moduli spaces of stable pairs', *Int. J. Math.* **6** 893, alg-geom/9505012; (1996). 'Quaternionic monopoles', *Commun. Math. Phys.* **180** 363, alg-geom/9505029.

[63] S. Bradlow and O. García-Prada. (1997). 'Non-abelian monopoles and vortices', *Lect. Notes Pure Appl. Math.*, **184**, 567.

[64] P. Feehan and T. Lenness. (1998). 'PU(2) monopoles and relations between four-manifold invariants', *Topology Appl.*, **88**, 111; (1998). 'PU(2) monopoles, I: Regularity, Uhlenbeck compactness, and transversality', *J. Diff. Geom.*, **49**, 265; (2001). 'PU(2) monopoles, II: Top-level Seiberg-Witten moduli spaces and Witten's conjecture in low degrees', *J. Reine Angew. Math.*, **538**, 135; (2001). 'PU(2) monopoles and links of top-level Seiberg-Witten moduli spaces', *J. Reine Angew. Math.*, **538**, 57; (2001). 'On Donaldson and Seiberg-Witten invariants', math.DG/0106221.

[65] S. Kobayashi and K. Nomizu. (1996). *Foundations of differentiable geometry,* (Wiley).

[66] R. Bott and L. Tu. (1982). *Differential forms in algebraic topology,* (Springer-Verlag).

[67] E. Witten. (1992). 'The N matrix model and gauged WZW models', *Nucl. Phys.*, **B 371**, 191.

[68] V. Mathai and D. Quillen. (1986). 'Superconnections, Thom classes, and equivariant differential forms', *Topology*, **25**, 85.

[69] M.F. Atiyah and L. Jeffrey. (1990). 'Topological Lagrangians and cohomology', *J. Geom. Phys.* , **7**, 119.

[70] M. Blau. (1993). 'The Mathai-Quillen formalism and topological field theory', *J. Geom. Phys.*, **11**, 95, hep-th/9203026.

[71] M. Blau and G. Thompson. (1995). 'Localization and diagonalization: A review of functional integral techniques for low dimensional gauge

theories and topological field theories', *J. Math. Phys.*, **36**, 2192, hep-th/9501075.

[72] C. Becchi, A. Rouet and R. Stora. (1976). 'Renormalization of gauge theories', *Ann. Phys.*, **98**, 287.

[73] I.V. Tyupin. (1975). 'Gauge invariance in field theory and statistical physics in the operator formalism', Lebedev preprint FIAN No. 39 (in Russian), unpublished.

[74] M. Henneaux. (1985). 'Hamiltonian form of the path integral for theories with gauge freedom', *Phys. Rept.*, **126**, 1.

[75] L. Alvarez-Gaumé. (1983). 'Supersymmetry And The Atiyah-Singer Index Theorem', *Commun. Math. Phys.*, **90**, 161; (1985). 'Supersymmetry and index theory' in *Supersymmetry*, (eds. K. Dietz et al.), NATO Adv. Study Institute, (Plenum, New York).

[76] D. Friedan and P. Windey. (1984). 'Supersymmetric derivation of the Atiyah-Singer index and the chiral anomaly', *Nucl. Phys.*, **B235**, 395.

[77] E. Witten. (1992). 'Mirror Manifolds and Topological Field Theory', in *Essays on Mirror Manifolds*, ed. S.-T. Yau (International Press).

[78] J.M.F. Labastida and P.M. Llatas. (1992). 'Topological matter in two dimensions', *Nucl. Phys.*, **B379**, 220.

[79] L. Álvarez-Gaumé and D.Z. Freedman. (1983). 'Potentials for the Supersymmetric Nonlinear Sigma Model', *Commun. Math. Phys.*, **91**, 87.

[80] J.M.F. Labastida and P.M. Llatas. (1991). 'Potentials for topological sigma models', *Phys. Lett.*, **B 271**, 101.

[81] J.M.F. Labastida and M. Mariño. (1997). 'Twisted $\mathcal{N} = 2$ supersymmetry with central charge and equivariant cohomology', *Comm. Math. Phys.*, **185**, 37, hep-th/9603169.

[82] L. Álvarez-Gaumé and F. Zamora. (1998). 'Lectures in topological quantum field theory', (eds. H. Falomir, R. Gamboa and F. Schaposnik), Proceedings of the CERN-Santiago de Compostela-La Plata Meeting on *Trends in Theoretical Physics*, (American Institute of Physics, New York), CP 419, 1; hep-th/9709180.

[83] M. Peskin. (1997). 'Duality in supersymmetric Yang–Mills theory', hep-th/9702094.

[84] W. Lerche. (1997). 'Introduction to Seiberg-Witten theory and its stringy origin', *Nucl. Phys. Proc. Suppl.*, **55B**, 83-117, hep-th/9611190.

[85] Nathan Seiberg. (1993). 'Naturalness versus supersymmetric non-

renormalization theorems', *Phys. Lett.*, **B318**, 469, hep-ph/9309335; (1994). 'Exact results on the space of vacua of four-dimensional SUSY gauge theories', *Phys. Rev.*, **D49**, 6857, hep-th/9402044; (1994). 'The power of holomorphy', hep-th/9408013.

[86] N. Seiberg. (1988). 'Supersymmetry and non-perturbative beta functions', *Phys. Lett.*, **B206**, 75.

[87] G. 't Hooft. (1976). 'Computation of the quantum effects due to a four-dimensional pseudoparticle', *Phys. Rev.*, **D 14**, 3432.

[88] N. Dorey, T.J. Hollowood, V.V. Khoze and M.P. Mattis. (2002). 'The calculus of many instantons', *Phys. Rept.*, **371**, 231, hep-th/0206063.

[89] N. Nekrasov. (2004). 'Seiberg-Witten prepotential from instanton counting', *Adv. Theor. Math. Phys.*, **7**, 831, hep-th/0206161.

[90] N. Nekrasov and A. Okounkov. (2003). 'Seiberg-Witten theory and random partitions', hep-th/0306238.

[91] K. Fujikawa. (1980). 'Path integral measure for gauge invariant fermion theories', *Phys. Rev.*, **D21**, 2848.

[92] M. Matone. (1995). 'Instantons and recursion relations in $\mathcal{N} = 2$ supersymmetric gauge theory', *Phys. Lett.*, **B367**, 342; hep-th/9506102.

[93] E. Witten. (1979). 'Dyons of charge $e\theta/2\pi$', *Phys. Lett.*, **B86**, 283.

[94] M. Di Pierro and K. Konishi. (1996). 'Mass, confinement and CP invariance in the Seiberg-Witten model', *Phys. Lett.*, **B388**, 90, hep-th/9605178.

[95] N.I. Akhiezer. (1990). *Elements of the theory of elliptic functions*, (American Mathematical Society).

[96] N. Koblitz. (1993). *Introduction to elliptic curves and modular forms*, second edition, (Springer-Verlag).

[97] L. Álvarez-Gaumé, M. Mariño and F. Zamora. (1998). 'Softly broken $\mathcal{N} = 2$ QCD with massive quark hypermultiplets', *Int. J. Mod. Phys.*, **A13**, 1847, hep-th/9703072.

[98] A. Bilal and F. Ferrari. (1998). 'The BPS spectra and superconformal points in massive $\mathcal{N} = 2$ supersymmetric QCD', *Nucl. Phys.*, **B516**, 175, hep-th/9706145; F. Ferrari. (1997). 'The dyon spectra of finite gauge theories', *Nucl. Phys.*, **B501**, 53, hep-th/9702166.

[99] H. Kanno and S.-K. Yang. (1998). 'Donaldson–Witten functions of massless $\mathcal{N} = 2$ supersymmetric QCD', *Nucl. Phys.*, **B 535**, 512; hep-th/9806015.

[100] M. Mariño and G. Moore. (1998). 'Integrating over the Coulomb branch in $\mathcal{N} = 2$ gauge theory', *Nucl. Phys. Proc. Suppl.*, **B68**, 336, hep-th/9712062.

[101] P. Deligne et al. (1999). *Quantum fields and strings, a course for mathematicians*, (American Mathematical Society).

[102] A. Losev, N. Nekrasov and S. Shatashvili. (1998). 'Issues in topological gauge theory', *Nucl. Phys.*, **B534**, 549, hep-th/9711108. (1997). 'Testing Seiberg-Witten solution', in *Strings, Branes, and Dualities*, Cargese Summer School, hep-th/9801061.

[103] M. Mariño and G. Moore. (1999). 'Donaldson invariants for non-simply connected manifolds', *Commun. Math. Phys.*, **203**, 249, hep-th/9804104.

[104] C. Lozano and M. Mariño. (2001). 'Donaldson invariants of product ruled surfaces and two-dimensional gauge theories', *Commun. Math. Phys.*, **220**, 231, hep-th/9907165.

[105] C. Vafa and E. Witten. (1994). 'A strong coupling test of S-duality," *Nucl. Phys.*, **B431**, 3, hep-th/9408074

[106] E. Witten. (1995). 'On S-duality in abelian gauge theory', *Selecta Mathematica, New Series*, **1**, 383, hep-th/9505186.

[107] E. Witten. (1994). 'Supersymmetric Yang–Mills theory on a four-manifold', *J. Math. Phys.*, **35**, 5101, hep-th/9403193.

[108] M. Mariño and G. Moore. (1999). 'The Donaldson–Witten function for gauge groups of rank larger than one', *Commun. Math. Phys.*, **199**, 25, hep-th/9802185.

[109] V. Pidstrigach and A. Tyurin. (1995). 'Localisation of the Donaldson invariants along Seiberg-Witten classes', dg-ga/9507004.

[110] R. Fintushel and R. Stern. (1996). 'The blow-up formula for Donaldson invariants', *Ann. of Math.*, **143**, 529; alg-geom/9405002.

[111] R. Fintushel and R. Stern. (1995). 'Immersed spheres in 4-manifolds and the immersed Thom conjecture' *Turk. J. Math.*, **19**, 145.

[112] L. Dixon, V. Kaplunovsky and J. Louis. (1991). 'Moduli dependence of string loop corrections to gauge coupling constants', *Nucl. Phys.*, **B355**, 649.

[113] J.A. Harvey and G. Moore. (1996). 'Algebras, BPS states, and strings', *Nucl. Phys.*, **B463**, 315, hep-th/9510182.

[114] R.E. Borcherds. (1998). 'Automorphic forms with singularities on Grassmannians', *Invent. Math.*, **132**, 491.

[115] L. Göttsche and D. Zagier. (1998). 'Jacobi forms and the structure of Donaldson invariants for 4-manifolds with $b_+ = 1$', *Sel. Math.*, New ser. **4**, 69, alg-geom/9612020.

[116] A. Losev, N. Nekrasov and S. Shatashvili. (2000). 'Freckled instantons in two and four dimensions', *Class. Quant. Grav.*, **17**, 1181-1187, hep-th/9911099.

[117] M. Bershadsky, A. Johansen, V. Sadov, and C. Vafa. (1995). 'Topological Reduction of 4D SYM to 2D σ–Models', *Nucl. Phys.*, **B448**, 166, hep-th/9501096.

[118] M. Thaddeus. (1997). 'An introduction to the moduli space of stable bundles on Riemann surfaces', *Lect. Notes Pure Appl. Math.*, **184**, 71.

[119] M. Thaddeus, 'Conformal field theory and the cohomology of the moduli space of stable bundles', *J. Diff. Geom.*, **35**, 131.

[120] V. Muñoz. (1999). 'Quantum cohomology of the moduli space of stable bundles over a Riemann surface', *Duke Math. J.*, **98**, 525, alg-geom/9711030.

[121] E. Witten. (1992). 'Two-dimensional gauge theories revisited', *J. Geom. Phys.*, **9**, 303; hep-th/9204083.

[122] M. Blau and G. Thompson. (1994). 'Lectures on 2d gauge theories', in *1993 Trieste Summer School in High Energy Physics and Cosmology*, (eds. E. Gava et al.), (World Scientific), hep-th/9310144.

[123] J. Morgan and Z. Szabó. (1999). 'Embedded tori in four-manifolds', *Topology*, **38**, 479.

[124] V. Muñoz. (1996) *Gauge theory and complex manifolds*, Ph. D. Thesis.

[125] J.M.F. Labastida and M. Mariño. (1995). 'Polynomial invariants for $SU(2)$ monopoles', *Nucl. Phys.*, **B 456**, 633,; hep-th/9507140.

[126] J.M.F. Labastida and C. Lozano. (1999). 'Duality in twisted $\mathcal{N} = 4$ supersymmetric gauge theories in four dimensions', *Nucl. Phys.*, **B537**, 203.

[127] P.C. Argyres and M.R. Douglas. (1995). 'New phenomena in $SU(3)$ supersymmetric gauge theory', *Nucl. Phys.*, **B448**, 93, hep-th/9505062.

[128] P.C. Argyres, M.R. Plesser, N. Seiberg, and E. Witten. (1996). 'New $\mathcal{N} = 2$ superconformal field theories in four dimensions', *Nucl. Phys.* **B461**, 71, hep-th/9511154.

[129] M. Mariño, G. Moore and G. Peradze. (1999). 'Superconformal invariance and the geography of four-manifolds', *Commun. Math. Phys.*, **205**, 691,

hep-th/9812055.

[130] M. Mariño, G. Moore and G. Peradze. (1999). 'Four-manifold geography and superconformal symmetry', *Math. Res. Lett.*, **6**, 429, math.DG/9812042.

[131] P.M.N. Feehan, P.B. Kronheimer, T.G. Leness, and T.S. Mrowka. (1999). '*PU*(2) monopoles and a conjecture of Mariño, Moore, and Peradze', *Math. Res. Lett.*, **6**, 169, math.DG/9812125.

[132] R. Fintushel and R.J. Stern. (2000). 'Nonsymplectic four-manifolds with one basic class', *Pac. J. Math.*, **194,** 325, math.GT/9907179.

[133] R. Fintushel J. Park, and R.J. Stern. (2002). 'Rational surfaces and symplectic 4-manifolds with one basic class', *Algebr. Geom. Topol.*, **2**, 391, math.GT/0202195.

[134] P.C. Argyres and A.E. Faraggi. (1995). 'Vacuum structure and spectrum of $\mathcal{N} = 2$ supersymmetric $SU(N)$ gauge theory', *Phys. Rev. Lett.*, **74**, 3931, hep-th/9411057.

[135] A. Klemm, W. Lerche, S. Theisen and S. Yankielowicz. (1995). 'Simple singularities and $\mathcal{N} = 2$ supersymmetric Yang–Mills theory', *Phys. Lett.*, **B344**, 169, hep-th/9411048.

[136] A. Klemm, W. Lerche and S. Theisen. (1996). 'Nonperturbative effective actions of $\mathcal{N} = 2$ supersymmetric gauge theories', *Int. J. Mod. Phys.*, **A10**, 1929, hep-th/9505150.

[137] M.R. Douglas and S.H. Shenker. (1995). 'Dynamics of $SU(N)$ supersymmetric gauge theory', *Nucl. Phys.*, **B447**, 271, hep-th/9503163.

[138] M. Mariño. (1999). 'The uses of Whitham hierarchies', *Prog. Theor. Phys. Suppl.*, **135**, 29, hep-th/9905053.

[139] A. Marshakov. (1999) *Seiberg-Witten theory and integrable systems*, (World Scientific).

[140] E. D'Hoker and D.H. Phong, 'Lectures on supersymmetric Yang–Mills theory and integrable systems', hep-th/9912271.

[141] J.D. Edelstein, M. Gómez-Reino and M. Mariño. (2000). 'Blowup formulae in Donaldson–Witten theory and integrable hierarchies', *Theor. Math. Phys.*, **4**, 503, hep-th/0006113.

[142] J.D. Edelstein and M. Gómez-Reino. (2000). 'Integrable hierarchies in Donaldson–Witten and Seiberg-Witten theories', in *Chicago 2000, Integrable hierarchies and modern physical theories*, p. 15, hep-th/0010061.

Mathematical Physics Studies

1. F.A.E. Pirani, D.C. Robinson and W.F. Shadwick: *Local Jet Bundle Formulation of Bäcklund Transformations*. 1979   ISBN 90-277-1036-8
2. W.O. Amrein: *Non-Relativistic Quantum Dynamics*. 1981
                                                      ISBN 90-277-1324-3
3. M. Cahen, M. de Wilde, L. Lemaire and L. Vanhecke (eds.): *Differential Geometry and Mathematical Physics*. 1983   Pb ISBN 90-277-1508-4
4. A.O. Barut (ed.): *Quantum Theory, Groups, Fields and Particles*. 1983
                                                      ISBN 90-277-1552-1
5. G. Lindblad: *Non-Equilibrium Entropy and Irreversibility*. 1983
                                                      ISBN 90-277-1640-4
6. S. Sternberg (ed.): *Differential Geometric Methods in Mathematical Physics*. 1984   ISBN 90-277-1781-8
7. J.P. Jurzak: *Unbounded Non-Commutative Integration*. 1985
                                                      ISBN 90-277-1815-6
8. C. Fronsdal (ed.): *Essays on Supersymmetry*. 1986   ISBN 90-277-2207-2
9. V.N. Popov and V.S. Yarunin: *Collective Effects in Quantum Statistics of Radiation and Matter*. 1988   ISBN 90-277-2735-X
10. M. Cahen and M. Flato (eds.): *Quantum Theories and Geometry*. 1988
                                                      ISBN 90-277-2803-8
11. Bernard Prum and Jean Claude Fort: *Processes on a Lattice and Gibbs Measures*. 1991   ISBN 0-7923-1069-1
12. A. Boutet de Monvel, Petre Dita, Gheorghe Nenciu and Radu Purice (eds.): *Recent Developments in Quantum Mechanics*. 1991   ISBN 0-7923-1148-5
13. R. Gielerak, J. Lukierski and Z. Popowicz (eds.): *Quantum Groups and Related Topics*. Proceedings of the First Max Born Symposium. 1992
                                                      ISBN 0-7923-1924-9
14. A. Lichnerowicz, *Magnetohydrodynamics: Waves and Shock Waves in Curved Space-Time*. 1994   ISBN 0-7923-2805-1
15. M. Flato, R. Kerner and A. Lichnerowicz (eds.): *Physics on Manifolds*. 1993
                                                      ISBN 0-7923-2500-1
16. H. Araki, K.R. Ito, A. Kishimoto and I. Ojima (eds.): *Quantum and Non-Commutative Analysis*. Past, Present and Future Perspectives. 1993
                                                      ISBN 0-7923-2532-X
17. D.Ya. Petrina: *Mathematical Foundations of Quantum Statistical Mechanics*. Continuous Systems. 1995   ISBN 0-7923-3258-X
18. J. Bertrand, M. Flato, J.-P. Gazeau, M. Irac-Astaud and D. Sternheimer (eds.): *Modern Group Theoretical Methods in Physics*. Proceedings of the Conference in honour of Guy Rideau. 1995   ISBN 0-7923-3645-3
19. A. Boutet de Monvel and V. Marchenko (eds.): *Algebraic and Geometric Methods in Mathematical Physics*. Proceedings of the Kaciveli Summer School, Crimea, Ukraine, 1993. 1996   ISBN 0-7923-3909-6

Mathematical Physics Studies

20.   D. Sternheimer, J. Rawnsley and S. Gutt (eds.): *Deformation Theory and Symplectic Geometry*. Proceedings of the Ascona Meeting, June 1996. 1997
ISBN 0-7923-4525-8

21.   G. Dito and D. Sternheimer (eds.): *Conférence Moshé Flato 1999*. Quantization, Deformations, and Symmetries, Volume I. 2000
ISBN 0-7923-6540-2 / Set: 0-7923-6542-9

22.   G. Dito and D. Sternheimer (eds.): *Conférence Moshé Flato 1999*. Quantization, Deformations, and Symmetries, Volume II. 2000
ISBN 0-7923-6541-0 / Set: 0-7923-6542-9

23.   Y. Maeda, H. Moriyoshi, H. Omori, D. Sternheimer, T. Tate and S. Watamura (eds.): *Noncommutative Differential Geometry and Its Applications to Physics*. Proceedings of the Workshop at Shonan, Japan, June 1999. 2001
ISBN 0-7923-6930-0

24.   M. de Gosson (ed.): *Jean Leray '99 Conference Proceedings*. The Karlskrona Conference in Honor of Jean Leray. 2003   ISBN 1-4020-1378-7

25.   J. Labastida and M. Marino: *Topological Quartum Field Theory and Four Manifolds*. 2005   ISBN 1-4020-3058-4

26.   C. Gu, H. Hu and Z. Zhou: *Darboux Transformations in Integrable Systems*. Theory and their Applications to Geometry. 2004   ISBN 1-4020-3087-8